Liaison Functions 2019

Held at the 2019 AIChE Annual Meeting

Orlando, Florida, USA
10 - 15 November 2019

ISBN: 978-1-7138-0540-3

Printed from e-media with permission by:

Curran Associates, Inc.
57 Morehouse Lane
Red Hook, NY 12571

Some format issues inherent in the e-media version may also appear in this print version.

Printed with permission by Curran Associates, Inc. (2020)

For permission requests, please contact AIChE
at the address below.

AIChE
120 Wall Street, FL 23
New York, NY 10005-4020

Phone: (800) 242-4363
Fax: (203) 775-5177

www.aiche.org

Additional copies of this publication are available from:

Curran Associates, Inc.
57 Morehouse Lane
Red Hook, NY 12571 USA
Phone: 845-758-0400
Fax: 845-758-2633
Email: curran@proceedings.com
Web: www.proceedings.com

TABLE OF CONTENTS

582663 RAPID Manufacturing Institute and AIChE

Process intensification is of worldwide interest as a type of advanced manufacturing. It aims for new, more compact, and more efficient processes, especially by applying component integration, novel driving forces, and modular deployment of these units. Join the PAIC Town Hall to hear about the leadership role AIChE plays via RAPID: a US Department of Energy advanced-manufacturing institute for process intensification and the development of process and operation principles and their applications.

582664 The Importance of Credentialing and Licensure

How do additional credentials strengthen an engineer's competitive position? What are the credentials that a chemical engineer may find helpful? How does a credential differ from additional education, licensure and advanced degrees? Information will be shared that will hopefully open up additional avenues for engineers of all ages to consider in their career planning.

582665 Legal Developments on Plastics: Basel Convention and the Circular Plastics Economy

The PAIC Legal Developments Subcommittee follows legal issues arising in: energy, food, water and other resources; climate change; chemical regulation; anti-terrorism, physical and cybersecurity, and chemical process safety; ethics, manufacturing and business law. At the PAIC Town Hall, we will review the May 2019 addition of plastic waste to the Basel Convention and its provisions regarding the development of a worldwide circular plastics economy; and the progress of the Expert Working Group pursuant to the Convention.

582666 Responding to Paradigm Shifts

Paradigm shifts generally lead to winners and losers. Manufacturing, chemicals, and energy illustrate three important areas of change driven by technology innovation and a shift from product to integrated system thinking. At the PAIC Town Hall, we will discuss the need for new frameworks that mesh technology, societal, and environmental dimensions and for new ways to respond to on-going shifts.

Preparing for the Worst:

The Case for Solar Geoengineering Research and Oversight

2019

AUGUST 6

Bradie S. Crandall
American Institute of Chemical Engineers

AIChE The Global Home of Chemical Engineers

UNIVERSITY OF South Carolina

WISE

"The Earth is the only world known so far to harbor life. There is nowhere else, at least in the near future, to which our species could migrate. Visit, yes. Settle, not yet. Like it or not, for the moment the Earth is where we make our stand."

-Carl Sagan

Table of Contents

Executive Summary

As global carbon dioxide emissions continue to rise, the looming threat of a climate crisis is becoming a growing concern. To avoid the intensification of geo-political, health, economic, and environmental risks, the world may begin considering more drastic measures to address climate change. One of these drastic measures that has been discussed is solar geoengineering. This strategy involves taking measures to intervene with the earth's climate by modifying the planet's albedo. A variety of key conflicts and concerns related to the prospect of solar geoengineering will be considered in this work. These concerns include the potential of a research echo chamber that may occur in this esoteric field which could amplify false notions, the existence of a fine line between solar geoengineering research and implementation, a lack of understanding of the risks associated with this technology, and the possibility of rogue actors implementing solar geoengineering unilaterally. Additionally, the issue of termination shock (if solar geoengineering efforts were to abruptly stop following implementation) will be addressed along with the possible conflation of the climatic effects of climate change with the effects of solar geoengineering. Finally, the lack of public understanding about solar geoengineering technologies and the moral hazard of replacing efforts to reduce emissions with solar geoengineering efforts will be discussed.

Policy recommendations to address the growing interest in solar geoengineering include mechanisms to prioritize climate change mitigation and adaptation, to promote federally funded research, to establish international oversight, to encourage transparency, and to fill the legal void. It is critical for the U.S. to prioritize climate change mitigation and adaptation strategies prior to advancing federal solar geoengineering efforts to avoid being accused of a moral hazard. This prioritization should be accomplished by rejoining the Paris Agreement as soon as possible and enacting policy to achieve the goals established by the international treaty. A road map will be needed to generate robust climate change policies that will strive to attain the Paris Agreement goals. It is suggested that a bipartisan "Green Deal" framework consisting of the commonalities between the Green New Deal and Green Real Deal be constructed. If these efforts fail to address climate change and solar geoengineering is needed, the establishment of international oversight is critical to ensure multilateral deployment. It is recommended that the previously established United Nations Intergovernmental Panel on Climate Change (IPCC) perform annual risk evaluations on the implementation of solar geoengineering. It is also recommended that a U.S. Federal Solar Geoengineering Research Program be created to gain a better understanding of this emerging technology. The DOE's Office of Science should oversee this program by providing research grants to universities and national laboratories. Additionally, NASA and NOAA would provide their satellite resources to aid research efforts while the EPA would provide environmental impact assessments to ensure all environmental risks are minimized during experimentation. It is critical for solar geoengineering research to be transparent to prevent public backlash. This should be accomplished via improved public education and the establishment of national and international research clearinghouses. Finally, the solar geoengineering legal void should be filled by outlawing unilateral action outside of the IPCC's oversight framework and expanding international environmental impact assessment law.

Foreword

About the Author

Bradie S. Crandall is a senior at the University of South Carolina and will graduate in May 2020 with a B.S. in Chemical Engineering, a minor in Chemistry, and with leadership distinction in research. Upon graduating, he intends to further his education by pursuing a PhD in Chemical Engineering. His interests include sustainable energy production, emission reduction, climate change mitigation policy, and science education. He currently serves as the treasurer of the University of South Carolina's student chapter of the American Institute of Chemical Engineers. He also works to promote inclusivity in the field of engineering as a member of the Society of Women in Engineering. He is a recipient of the Magellan Scholar Grant and was appointed to Oak Ridge National Laboratory by the Department of Energy in 2018. He has over three years of environmental catalysis experience at the University of South Carolina's Strategic Approaches to the Generation of Electricity Research Center. His research projects have included work on the tri-reforming of methane, adsorptive desulfurization of liquid fuels, and photocatalytic hydrogen production via water splitting.

About the WISE Program

The Washington Internships for Students of Engineering (WISE) program was founded in 1980 through the collaborative efforts of several professional engineering societies and has become one of the premier Washington internship programs. Each summer, the WISE societies select outstanding 3rd or 4th year engineering/computer science students, or students in engineering/computer science graduate programs, from a nation-wide pool of applicants. The students spend nine weeks living in Washington, D.C. during which they learn how government officials make decisions on complex technological issues, and how engineers can contribute to the legislative process and regulatory public policy decision-making.

Acknowledgements

First and foremost, I would like to express the sincerest of gratitude to my parents, Brad and Jodie Crandall. Without them, I would not be the man I am today. The sincerest of thanks is also given to my longtime girlfriend, Rachel Margolis, for supporting me in all my endeavors. I would like to recognize the American Institute of Chemical Engineers for their sponsorship and aid this summer with a special thanks to Dr. Dale Keairns and Heather Yuengling. I owe the opportunity to have such a positive experience in DC to the Institute of Electrical and Electronics Engineers and I would like to especially thank Diana Librizzi and Erica Wissolik for hosting the other interns and I this summer. All my professors at the University of South Carolina, especially Dr. John Weidner, Dr. Johen Lauterbach, and Dr. Erdem Sasmaz should be recognized for imparting their engineering knowledge upon me. I would also like to thank the interns whom I worked alongside in DC this summer. Finally, I would have never become a chemical engineer had my high school chemistry teacher, Victor Senn, not inspired me to do so. May he rest in peace.

Acronyms

CCS – Carbon capture and sequestration

CDR – Carbon dioxide removal

CIA – Criminal Intelligence Agency

CTBT – Comprehensive Nuclear-Test-Ban Treaty

DOE – Department of Energy

EIA – Environmental Impact Assessment

EPA – Environmental Protection Agency

IPCC – Intergovernmental Panel on Climate Change

NAS – National Academy of Sciences

NASA – National Aeronautics and Space Administration

NOAA – National Oceanic and Atmospheric Administration

SCoPEx – Stratospheric Controlled Perturbation Experiment

SPICE – Stratospheric Particle Injection for Climate Change Engineering

SRM – Solar radiation management/modification

UNEP – United Nations Environment Programme

UN – United Nations

UV – Ultraviolet

1. Introduction

1.1 The Climate Crisis

The ability of human beings to modify the composition of earth's atmosphere on a global scale to an extent that it warms the planet via the greenhouse effect is now certain (*Fig. 1*). The amount of carbon dioxide in the atmosphere today is the highest it has been in 3 million years, before humans walked the earth [1]. Even more alarming, carbon dioxide levels in the atmosphere are increasing over 100 times faster than the increase observed at the end of the last ice age, more rapidly than has ever been recorded in earth's history [2]. In 2018, the United Nation's Intergovernmental Panel on Climate Change (IPCC) reported that the world has only until 2030 to avoid potentially irreversible climate change [3]. However, it has recently become clear that the models used by the IPCC to assess the impacts of climate change have historically underestimated warming projections [4,5].

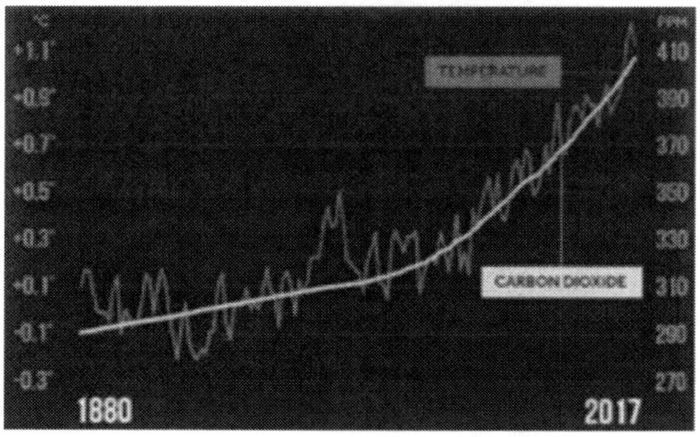

Figure 1. Human caused global temperature and carbon dioxide increase [6].

In 2018, the IPCC reduced the holding requirement for the temperature rise from 2 to 1.5 °C after realizing this was necessary to avoid some of the worst, most irreversible consequences of climate change [3]. Over a 1 °C increase in temperature has already occurred [6-7]. This new 1.5 °C holding temperature requires that global carbon dioxide emissions peak in 2020, are halved in 2030, and halved once again by 2040. Leading climatologists have predicted that global average temperature is likely to rise 5-6 °C by 2100 based upon current trends (fig. 2). In 2018, U.S. emissions rose by about 3% from the previous year and are expected to rise again in 2019 following President Trump's rescindment of President Obama's emission reduction initiatives [8-9]. Although not impossible, it is unlikely that U.S. carbon dioxide emissions will plateau in 2020 as needed to achieve the IPCC's 1.5 °C holding requirement.

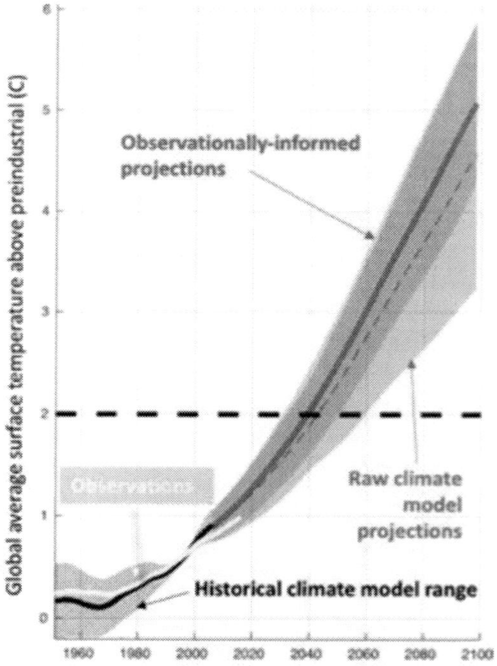

Figure 2. Comparison of raw and observationally-informed climate model projections [7].

The extent to which climate change will impact society remains unclear. However, sufficient evidence exists to identify the following outcomes: temperature rise, ocean acidification, sea level rise, loss of biodiversity, artic sea ice loss, increase in extreme weather events, increase in drought frequency, more heat waves, extreme changes in precipitation patterns, and increased forest fire activity [10]. These outcomes will have wide-ranging geopolitical consequences and will act as conflict intensifiers that pose serious threats to national security. Over the next 20 years, the effects of climate change have the potential to lead to global perturbations, increased risk of political instability, heightened tensions between countries for resources, a growing number of climate-linked humanitarian crises, emergent geostrategic competitive domains, and adverse effects on militaries [11].

The economic cost of climate change to the United States is incredibly difficult to quantify, but the Environmental Protection Agency (EPA) predicts it will exceed $500 billion annually [12]. This cost is likely a gross underestimation since a variety of factors including the potential for an outbreak of war, the true value of human life, and foreign humanitarian aid cannot be predicted. However, the EPA's estimation provides an indication that climate change will have a significant impact on the U.S. economy.

There is no longer a serious debate about the existence or the cause of the changes in the climate currently being experienced. Climate change is now here, and its effects are being observed across the globe in real time. Scientific consensus has concluded that human activity, specifically the emission of greenhouse gases, is the driving force behind climate change. The debate now lies in the hands of elected officials to decide how the United States will respond to this crisis.

1.2 Global Decarbonization Efforts

Under the looming threat of climate change, the developed world is currently undergoing a transition away from fossil fuels towards renewable energy sources. This transition is necessary to cut emissions and meet the targets established by the United Nations Paris Agreement in 2015 [13]. Although some nations are on pace to meet their proposed targets, many are not. Additionally, global carbon emissions are still on the rise as global energy demand continues to increase more rapidly than renewable energy infrastructure is being deployed.

As the undeveloped world pursues the standard of living enjoyed by the developed world, they are becoming consumers of coal and oil to meet their increased energy needs. The undeveloped world's energy demands will continue to grow with improved standards of living, further exacerbating climate change. Renewable energy technologies have failed to be implemented in the undeveloped world largely due to a lack of government funding for research and development and poor infrastructure. Although drastic improvements in affordability have been made in renewable energy technologies, deployment has been sparse in the undeveloped global south.

A significant amount of effort to decarbonize the energy sector has been put forth; however, agricultural, manufacturing, and transportation sectors will also need decarbonized to curb climate change. Efforts to decarbonize the energy sector at a rate greater than the rate at which energy demand is increasing have largely fallen short. Only drastic action to decarbonize the economy on a global scale within the next decade will prevent significant damage caused by climate change. At the current level of ambition, the world will continue suffer from unfavorable changes in the climate that will only worsen. If a technology emerged that could buy society more time to decarbonize in an economical fashion, the deployment of this technology should be seriously considered.

1.3 Solar Geoengineering

Geoengineering, also known as climate intervention, can be defined as the intentional modification of the earth's climate to mitigate climate change. Geoengineering should not be conflated with baseless chemtrail conspiracy theories. The field of geoengineering research is based strictly upon factual and scientific grounds. Geoengineering strategies fall into two categories, carbon dioxide removal (CDR) and solar radiation management (SRM). A significant amount of effort has recently been focused on CDR strategies such as carbon capture and sequestration (CCS) in the United States. CCS typically involves capturing carbon from large point sources; however, CCS can be employed as a CDR geoengineering strategy via direct air capture that scrubs carbon dioxide from the ambient atmosphere. Since CDR does not pose the same geopolitical issues and international relation challenges as SRM, the focus of this work is on the latter.

Solar radiation management, or albedo modification, relates to strategies designed to reduce the amount of solar radiation that contacts the surface of the earth. SRM does not affect the concentration of greenhouse gases in the atmosphere, it simply reduces global warming via solar reflection. The most commonly discussed SRM method involves the addition of aerosols to the lower stratosphere to reflect about 1% of incoming solar radiation.

SRM addresses the warming symptom of climate change but does not resolve all the changes in the climate caused by the release of greenhouse gas emissions. Therefore, some symptoms of climate change such as ocean acidification will not be affected by SRM [1]. It must be made clear that the potential for SRM implementation should not hinder efforts to reduce greenhouse gas emissions. SRM should be viewed as an extreme measure to be employed as a last resort to mitigate climate change. The feasibility and environmental impacts of SRM still remain unclear [14].

2. Background

2.1 History

There have been a variety of attempts scattered throughout human history to manipulate weather. However, these attempts should not be confused with geoengineering since they deal with weather rather than the climate. Solar geoengineering was first seriously considered in 1965 by President Lyndon B. Johnson's Science Advisory Committee in the first high-level government statement on global warming. This committee issued a report that considers, "deliberately bringing about countervailing climatic changes," by "raising the albedo, or reflectivity, of the Earth" [15]. Nearly a decade later, in 1974, Soviet climatologist Mikhail Budyko investigated the potential to reverse global warming by burning sulfur in the stratosphere to mimic the climatic effects of a volcano [16]. In 1976 the term "geoengineering" was first coined by Cesare Marchetti as a kind of system synthesis that took a global perspective to the carbon dioxide problem [17].

As the potential to avoid the consequences of climate change appeared less likely at the turn of the millennium, interest in solar geoengineering begins to surge. In 2006, Paul Crutzen, winner of the 1995 Nobel Prize in Chemistry for his ozone research, advocates for solar geoengineering over emission reductions [18]. This advocacy faced widespread criticism from the scientific community. In 2006, Pentagon weapons designer, Lowell Wood proposes a scheme to use artillery to launch sulfates into the Artic stratosphere to cool the planet at a NASA conference. 3 years later, while referring to geoengineering as one of the potential strategies to address climate change, President Obama's chief science adviser stated, "It's got to be looked at" [19].

The desire to begin conducting serious real-world solar geoengineering research begins to appear in 2011. The Stratospheric Particle Injection for Climate Change Engineering (SPICE), a British academic consortium, attempts to plan the first real-world solar geoengineering experiment [20]. However, this field test was called off in 2012 due to controversy caused by a lack of public engagement and transparency. That same year, The National Natural Science Foundation of China begins prioritizing geoengineering research [21]. In response, the Central Intelligence Agency (CIA) partners with the National Academy of Sciences (NAS) to begin performing an evaluation of SRM geoengineering techniques [22]. In 2014, solar geoengineering strategies were first considered by the UN in the IPCC's fifth assessment report [23]. The sum of the events over the past few decades have built the foundation for the research currently being conducted on solar geoengineering and will continue to provide valuable insights to this emerging technology.

2.2 Recent Interest

In 2019, the prospect of solar geoengineering was being seriously considered as an option to mitigate climate change. In March 2019, Switzerland and nine other nations requested that the United Nations Environment Programme (UNEP) open an inquiry into potential geoengineering governance options [24]. The United States, along with Brazil and Saudi Arabia, blocked this resolution. However, geoengineering will be a key part of the IPCC's Sixth Assessment Report to be published in 2021.

In the United States, the NAS will be issuing a report in the first half of 2020 on solar geoengineering that will build upon their previous 2015 report. The motivation of expanding this report is based upon the lack of time remaining for humanity to mitigate climate change and the reality that climate intervention strategies may need to be considered in the future [25]. Additionally, there are a handful of research groups at universities that are dedicated to addressing the technological, socio-economic, and political challenges related to geoengineering. These research groups include Harvard's Solar Geoengineering Research Program and the Carnegie Climate Governance Initiative.

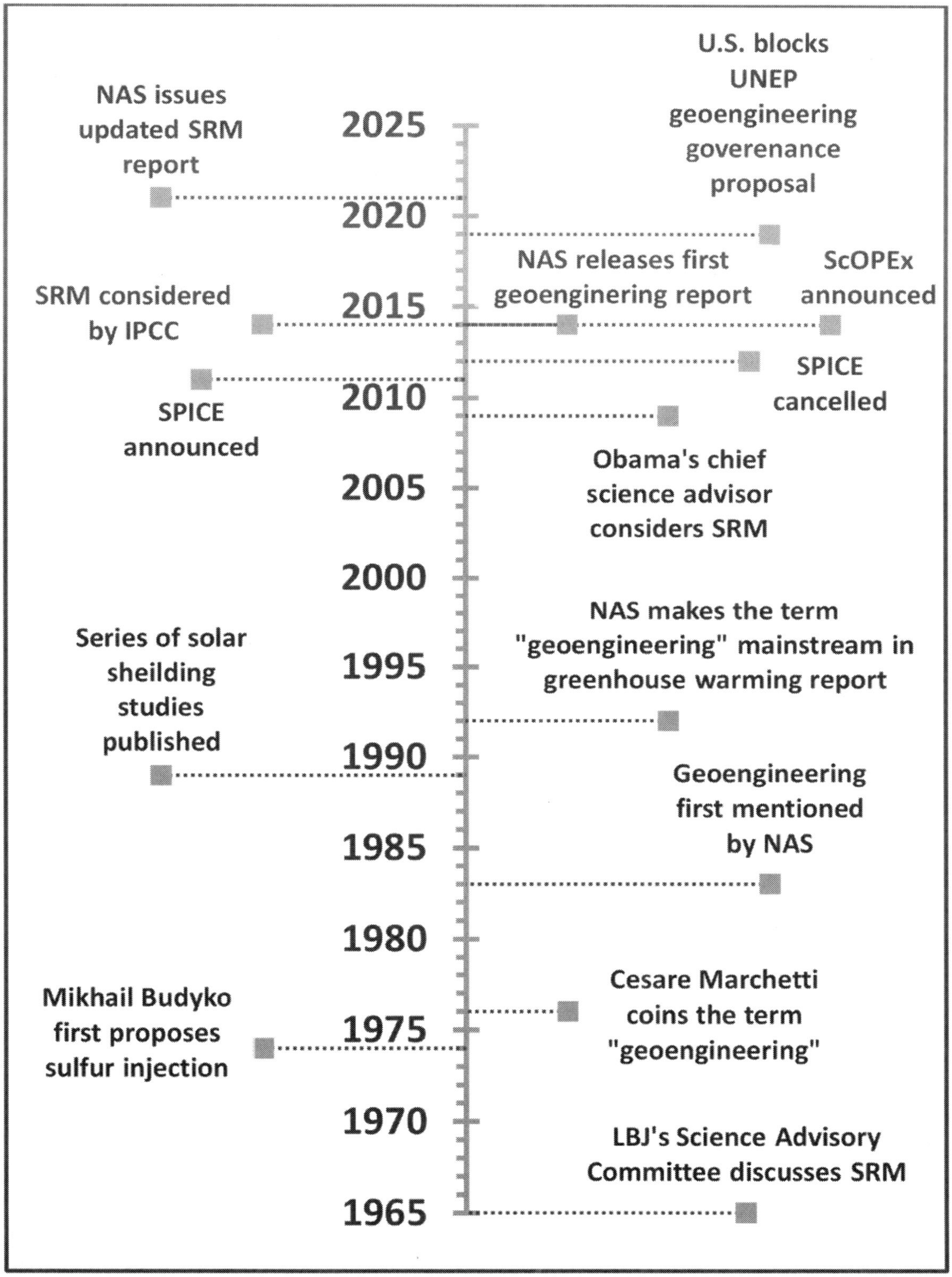

Figure 3. Timeline of geoengineering research and interest.

2.3 Technological Readiness and Feasibility

It is important to consider not only how ready solar geoengineering technology is for implementation, but also how practical implementation would be. The degree to which each solar geoengineering strategy has satisfied the issues of technological readiness and feasibility varies widely. The most prominent solar geoengineering strategies can be placed into four categories:

(1) **Stratospheric Aerosol Injection** – Reducing temperatures by injecting a reflective aerosol into the lower stratosphere

(2) **Marine Cloud Brightening** – Whitening clouds on land or seeding clouds above the ocean to increase planetary albedo (reflectivity)

(3) **Cirrus Thinning** – Decreasing the resistance to the escape of infrared radiation from the earth by thinning cirrus clouds

(4) **Surface Albedo Modification** – Brightening earth's surface (land or ocean) to increase reflectivity

Stratospheric Aerosol Injection appears to be the most promising solar geoengineering approach in terms of practicality and cost. By some estimates, the cost of this strategy could be as low as $1 billion/year [26]. Most recent research on solar geoengineering has been focused on the stratospheric aerosol injection approach. Stratospheric aerosol injection research has been purely based upon computer models and naturally occurring analogues, such as volcanoes, thus far. However, field tests may begin as soon as 2020. Harvard University's Keutsch Research Group is currently in the process of planning what has been dubbed the Stratospheric Controlled Perturbation Experiment (SCoPEx). This first of its kind experiment involves the injection of up to one kilogram of calcium carbonate into the stratosphere. Although most focus in stratospheric aerosol injection research has been on sulfates [27-31], calcium carbonate was chosen for the first trial of this experiment since it is a naturally occurring mineral dust. "SCoPEx will address questions about how particles interact with one another, with the background stratospheric air, and with solar and infrared radiation" [32]. This experiment will likely provide valuable information about the implementation of stratospheric aerosol injection and could improve the current models.

Marine Cloud Brightening research, like stratospheric aerosol injection research, has solely consisted of computer simulations and observations of naturally occurring phenomena. The general idea behind marine cloud brightening is to seed stratocumulus clouds with small seawater particles to enhance the clouds' albedo [33]. Marine cloud brightening has low technological readiness due to the historical lack of success associated with cloud seeding [34].

Cirrus Thinning is still technologically theoretical with most research being conducted via computer simulations. Cirrus clouds differ from other types of clouds in that they have a net warming effect rather than a cooling effect since they trap heat. Cirrus clouds can be thinned by seeding them with aerosols that will stimulate ice formation causing the clouds' ability to trap heat to be diminished [35]. Like with stratospheric aerosol injection, the cloud seeding technology needed to accomplish cirrus cloud thinning is far from perfected. Additionally, there is some evidence suggesting that over-seeding cirrus clouds may lead to warming rather than the desired cooling effect [36].

Surface Albedo Modification involves strategies to increase the reflectivity of earth's surface. Examples of surface albedo modification strategies include brightening roads and roofs, planting high albedo crops, snow forest clearance, and artic ice coverings [37-40]. Surface albedo modification is the simplest and most technologically ready of the selected solar geoengineering technologies, but this does not imply it is the most feasible or that the consequences of implementation are understood. There are considerable economic challenges associated with each surface albedo modification strategy. Until these economic challenges can be overcome, surface albedo modification has relatively low feasibility.

3. Key Conflicts and Concerns

3.1 Research Echo Chamber

A key challenge to obtaining reliable estimates of the pros and cons of solar geoengineering is that most solar geoengineering research is currently conducted by a relatively small group of people. The conversations surrounding solar geoengineering are often dominated by individuals such as Dr. David Keith of Harvard's Solar Geoengineering Research Program. Although there is no reason to doubt Dr. Keith's expertise in this field, additional perspectives from researchers with a more diverse set of backgrounds would be more conducive to advancing solar geoengineering research. The promotion of science relies upon problems to be approached in unique ways. A diverse group of individuals tends to accomplish this best [41].

Another danger of relying upon a small group of individuals to produce quality science is the potential for an echo chamber effect. This effect is characterized by an amplification of an idea within a closed system. This becomes dangerous when a false idea enters the closed system and is amplified to an extent that it is accepted as fact and no longer questioned. Since the same authors appear on most academic publications on solar geoengineering, the probability of the reinforcement of a false idea is high. A larger, more diverse group of researchers can help mitigate the echo chamber effect by rigorously questioning scientific ideas before they are amplified and considered common knowledge in the field.

3.2 Research vs Implementation

There is a fine line between solar geoengineering research and implementation. It is challenging to precisely establish a clear boundary between solar geoengineering research and implementation so these two ideas may be best represented on a spectrum (fig. 4).

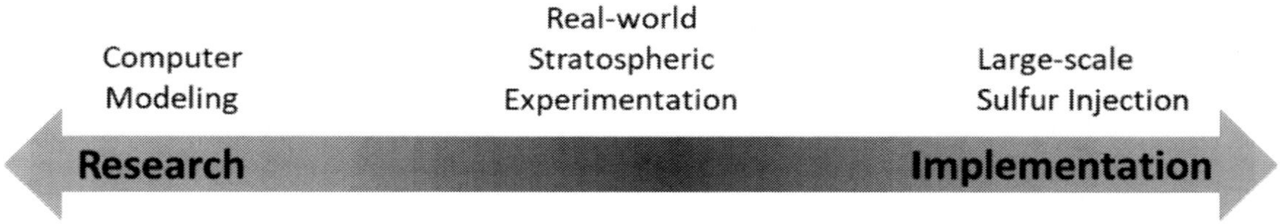

Figure 4. Solar Geoengineering Research/Implementation Spectrum

Solar geoengineering research that is conducted via computer modeling can clearly be categorized as research rather than implementation. However, real-world stratospheric experiments are much more difficult to categorize, and the categorization may vary from experiment to experiment. Small scale real world stratospheric experiments such as Harvard's SCoPEx likely fall more towards the research side of the spectrum than the implementation due to the innocuous nature of the experiment, but other experiments may fall closer to implementation.

Any real-world solar geoengineering experimentation must be performed very cautiously and on as small of a scale as possible to avoid crossing over into implementation territory. Stratospheric experimentation will inherently involve regional implementation, but global implementation must be avoided during such experiments. Real-world solar geoengineering experiments will be incredibly valuable to improving scientific understanding, but when probing uncharted territory, ensuring these experiments are low impact is critical to risk management.

3.3 Health and Environmental Risks

There is a plethora of unintended consequences associated with solar geoengineering implementation. Amongst the most concerning of these consequences are potential health and environmental risks. These risks are currently understood relatively poorly, but computer modeling has identified a handful of potential negative impacts worth considering. Specific consequences of these potential risks will be discussed further, but the potential risks of solar geoengineering can be summarized as:

- Changes in UV exposure and air quality
- Loss of tropical rainforest productivity
- Altered stratospheric circulation and jet stream destabilization
- Variations in ozone concentration
- Increase in methane lifetime
- Disproportionate temperature changes
- Deviations in natural precipitation patterns

Two of the possible negative implications of solar geoengineering are changes in ultraviolet (UV) exposure and air quality. One model has found that solar geoengineering may lead to a large stratospheric ozone increase that induces a significant reduction in surface UV-B irradiance [42]. UV-B irradiance is important since it is necessary for the body to naturally produce vitamin D, an essential nutrient for human life. This model has demonstrated that solar geoengineering may hinder the production of this nutrient possibly leading to exacerbated vitamin D deficiencies. Additionally, the model demonstrated a significant change in the composition of the troposphere. These changes were primarily driven by decreased humidity and a reduced UV-A and UV-B flux; however, it is difficult to predict how these tropospheric changes may further impact the earth's climate.

Another negative impact relates to the loss of tropical rainforest productivity. Two computer models have revealed that the gross primary productivity of tropical rainforests may decrease as a result of marine sky brightening [43]. This is troubling since tropical rainforests act as natural carbon sinks that aid in climate change mitigation by sequestering carbon from the air back into the earth. Further understanding of the impact that marine cloud brightening could have on tropical rainforests is necessary prior to considering implementation.

A third potential negative implication of solar geoengineering is a change in stratospheric circulation. One model has revealed that stratospheric aerosol injection may generate thermal winds in the lower stratosphere causing stratospheric vortices to intensify [44]. Since the troposphere and the stratosphere are intimately coupled, this shift in stratospheric circulation may lead to a shift in tropospheric jet streams. This shift is already occurring in the polar jet stream as a result of greenhouse gas emission driven climate change, but solar geoengineering may intensify this shift and cause further destabilization. A consequence of the destabilization of the polar jet stream is that warm air will travel further towards the pole and cold air will travel further towards the equator causing extreme temperature changes (fig 5).

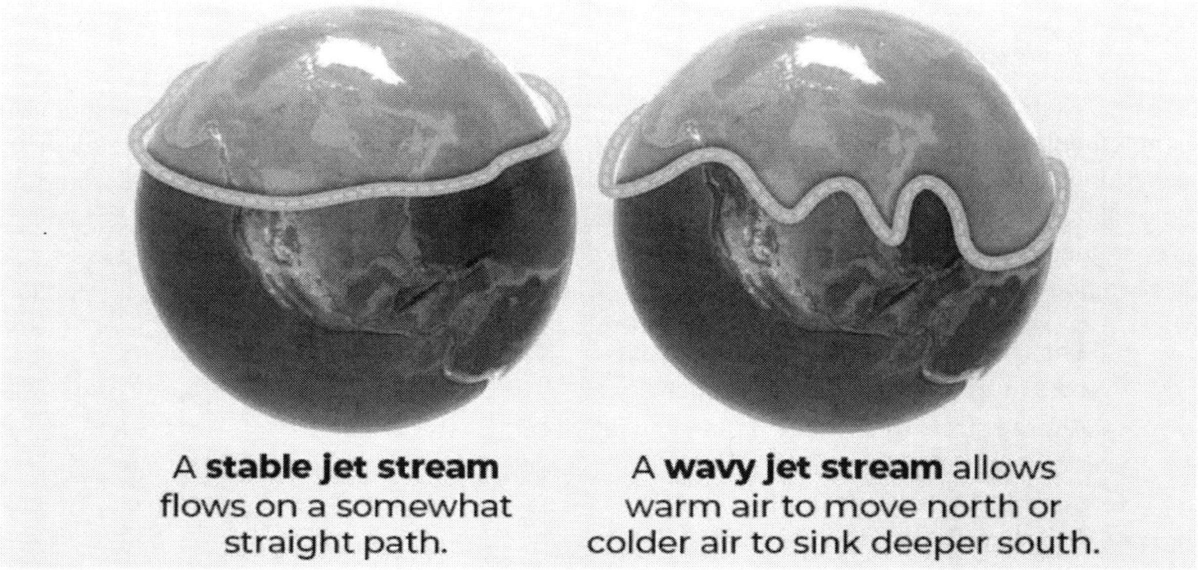

Figure 5. The Changing Jet Stream [45].

A fourth potential risk is the impact solar geoengineering may have on ocean life. A large portion of the world relies upon the ocean for food and solar geoengineering may decrease the net positive production of the ocean [46]. This could intensify the already worsening food shortages of the ocean caused by overfishing. Solar geoengineering may negatively impact net positive ocean production by changing physical factors, such as temperature, radiation, and ocean mixing; however, the effect of each of these individual factors is still relatively unclear.

A fifth risk associated with solar geoengineering are changes in ozone concentration. Balancing ozone concentration is important because if ground-level ozone becomes too concentrated, it can cause respiratory damage and harm crop production and vegetation. If ozone levels become too low, then more UV irradiation will reach earth's surface causing skin cancer rates to increase. One model has determined that solar geoengineering may either reduce or increase surface ozone levels depending upon which solar geoengineering strategy is implemented [47]. If stratospheric aerosol injection is deployed, then global surface ozone levels were modeled to decrease. If one of the other solar geoengineering strategies were deployed, then global surface ozone levels were modeled to increase. Stratospheric sulfates were found to decrease global surface ozone levels by disrupting the transport of stratospheric ozone to the troposphere via the disruption of natural photochemical reactions.

A sixth potential negative effect of solar geoengineering, specifically stratospheric aerosol injection, is an increase in methane lifetime. A computer simulation has found that the lifetime of methane, a potent greenhouse gas, may become significantly longer (approximately 16%) [28]. This is problematic since the global warming impact of methane is 28-36 times greater than that of carbon dioxide [48]. The potential of increasing the amount of time it takes for methane to naturally degrade in the atmosphere is an important consideration to make before implementing stratospheric aerosol injection.

In addition to a potential increase in methane lifetime, stratospheric aerosol injection may also lead to an extreme temperature and precipitation response. Two models have demonstrated that more extreme decreases in minimum land temperatures in comparison to minimum ocean temperatures may be experienced [49]. This shows that solar geoengineering implementation could lead to a disproportionate temperature decrease, creating the potential for winners and losers as a result of global solar geoengineering implementation. It was also found that maximum 5-day precipitation increases over subtropical oceans, warm spells decrease in the tropics, and the number of consecutive dry days decreases in the desert. A second study found an increase in peak precipitation duration over the subpolar climatic region, a decrease in peak precipitation duration over the tropical rain belt, and 1-3 month shift backwards in peak precipitation in much of the northern hemisphere [50]. This study illustrates the complexity of weighing the costs and benefits of solar geoengineering implementation.

A very limited amount of solar geoengineering research has been performed to gain an understanding of potential health and environmental consequences. Further research, beyond modeling work, is needed to better elucidate many potential problems. Additionally, many of the specific atmospheric mechanizations of the climate are poorly understood, further exacerbating the difficulty in quantifying the risks of solar geoengineering implementation. At the very least, it can be stated with near certainty that unintended health and environmental consequences will arise following implementation due to the difficulty of their prediction.

3.4 Rogue Actors

If the cost of solar geoengineering is as low as many researchers predict, the technology will be accessible to most somewhat developed nations and even a few wealthy individuals. This potential low cost, coupled with a lack of international laws restricting solar geoengineering, could enable rogue actors to act unilaterally to implement solar geoengineering. Unilateral deployment is highly undesirable given that earth's climate is a resource shared by all and should therefore, only be manipulated in a manner that maximizes the benefit for as many people as possible. If a single entity were to unilaterally implement solar geoengineering, it would likely be designed to solely benefit that individual rogue actor at the possible expense of other nations.

The probability that a rogue actor may attempt to deploy geoengineering is not negligible. Previous attempts have been made by rogue actors to geoengineer the climate. The most notable being the infamous case involving Palnktos, a Canadian "ocean restoration" startup. In July 2012, the director of Planktos, Russ George, dumped 100 tons of iron sulphate into the Pacific Ocean in an attempt to mitigate climate change. He claimed that the iron would stimulate algae growth that would remove additional carbon dioxide from the atmosphere via photosynthesis; however, this claim has been highly disputed by many experts [51]. Prior to the event, Canada's environment ministry warned George that the venture would violate international agreements [52]. Since international law is unequipped to deal with rogue actors, George's actions never got him into any legal trouble [53].

This legal void surrounding geoengineering efforts may encourage rogue actors to pursue unilateral implementation. The Planktos case has now set a dangerous precedent that there are no legal teeth in international law to grapple with rouge geoengineering. With the predicted low cost of geoengineering and the legal immunity precedent established, the chance of rogue actors unilaterally implementing solar geoengineering could be high.

3.5 Termination Shock

One of the most prominent concerns associated with solar geoengineering is termination shock. If solar geoengineering efforts were implemented over a long period of time and then suddenly stopped, a dangerous spike in global temperature could occur if emissions continue to rise after implementation [29, 54-57]. There are a variety of potential reasons the implementation of SRM may suddenly stop including war, a major shift in global politics, or if the cost were to become too high. The potential for termination shock should not be ruled out when implementation considerations are being made.

Although the impact that termination shock could have on the climate is unclear, it is incredibly likely that negative impacts may be experienced if atmospheric carbon dioxide concentrations continue to increase following solar geoengineering implementation. The most likely of these potential impacts is rapid global warming, but this may not be the only impact. Some of the other potential negative impacts that have been identified include an increase in global mean precipitation, decreased sea-ice cover, faster warming in high-latitudes and over land, and increased threats to biodiversity [54, 58].

In addition to the negative impacts, additional research is needed to determine how long it would take for termination shock to be experienced. One study estimates that 1-2 years after implementing stratospheric aerosol injection, the aerosol loadings would decrease and cause potential termination shock if additional aerosols were not injected within that time frame [54]. However, the maximum time of dormancy following the implementation of marine cloud brightening, cirrus thinning, or surface albedo modification solar geoengineering strategies before termination shock could occur remains unknown.

3.6 Climate Tug of War

Deciding where the global thermostat should be set via solar geoengineering is an incredibly difficult challenge. Due to the high probability of unintended consequences, it is likely that some nations could be unhappy with the effects of solar geoengineering implementation.

Many academics are considering the possibility that nations may choose to deploy counter-geoengineering efforts to reverse or neutralize solar geoengineering efforts, creating a climate "tug of war" [59-60]. Ultimately, a climate tug of war could lead to an oscillation between climate extremes causing a myriad of unintended consequences. Additionally, geoengineering efforts that are in opposition of each other is an extremely inefficient use of nations' resources.

Due to variation in geographic position of different nations, there may be disparity amongst climatic preferences. The environmental effects of climate change and their political and economic impacts are distributed unevenly globally. For example, Russia may gain access to oil and other natural resources as global warming causes the artic to melt, whereas African nations may experience crippling droughts [61-62]. Solar geoengineering efforts could be highly undesirable to Russia but may provide significant benefits to African nations. This is just a single example of the complications that must be considered to minimize the potential for counter-geoengineering efforts.

Additionally, counter geoengineering efforts have the potential to generate conflict between nations. This conflict may result in a climate war where nations compete to gain control of the climate via geoengineering. The Environmental Modification Convention, introduced on May 18th, 1977 in Geneva and ratified by the Unites States in 1980, bans the weaponization of weather [63]. However, it is unclear whether this treaty would apply to counter-geoengineering efforts.

3.7 Damage Attribution

Determining which climatic consequences are a direct result of solar geoengineering may prove to be challenging. Simple measurements of global average temperature would be an insufficient method to detect the effects of solar geoengineering. Moreover, it could take years to detect the effects solar geoengineering with a high level of confidence [64]. A wide variety of parameters would need to be measured to effectively determine the effects of solar geoengineering including global average temperature, precipitation, stratospheric circulation, UV exposure, and ozone concentration. Additionally, solid baselines of these parameters would need to be established prior to implementation. This may be difficult as many of these parameters have been rapidly changing due to anthropogenic climate change [23].

Another consideration that needs to be made is the potential for causal conflation between the effects of climate change and solar geoengineering following deployment of solar geoengineering. This issue could add another level of complexity to damage attribution. If solar geoengineering is implemented, it may be difficult to determine which consequences should be attributed to climate change and which should be attributed to the effects of solar geoengineering. This issue is also exacerbated by the difficulty associated with establishing baselines for climatic parameters. Error associated with poor baselines could compound error elsewhere since all measurements would be in comparison to this baseline. Furthermore, determining the amount of solar geoengineering to deploy in the future to mitigate damages following implementation becomes problematic due to the possibility of casual conflation. This issue must be addressed prior to the first implementation of solar geoengineering to ensure proper monitoring and verification of deployment.

3.8 Moral Hazard

Many commentators on solar geoengineering have expressed a fear that policy-makers and the fossil fuel industry may view this technology as an alternative to emission reduction efforts. It is well understood that the phenomena of risk compensation can arise when people respond to risk-reducing innovations by behaving less cautiously [65-70]. In the context of climate change, society may choose to emit additional greenhouse gases into the atmosphere if they feel that the risk of climate change has been reduced by solar geoengineering technology.

It has been argued that effective solar geoengineering may justify decreased mitigation efforts on the basis that a net decrease in global temperature may still be achieved without emission reductions [71]. The premise of this argument is flawed since it assumes that the only symptom of climate change that needs to be addressed is global warming. Solar geoengineering addresses the global warming symptom but fails to address other symptoms of climate change such as ocean acidification. The only effective strategy to mitigate ocean acidification is to reduce carbon emissions. Therefore, solar geoengineering implementation alone is not a strong justification to relax emission reduction efforts.

3.9 Public Reception

Prior to making any policy decisions related to solar geoengineering, the public reception of the technology must be understood. The primary issue related to the public reception of solar geoengineering is that most of the general public in the U.S. is unfamiliar with the technology (fig. 6). This is troubling given the likelihood that solar geoengineering discussions amongst policy-makers will begin in the near future, yet policy-makers lack information on the public perception of the issue.

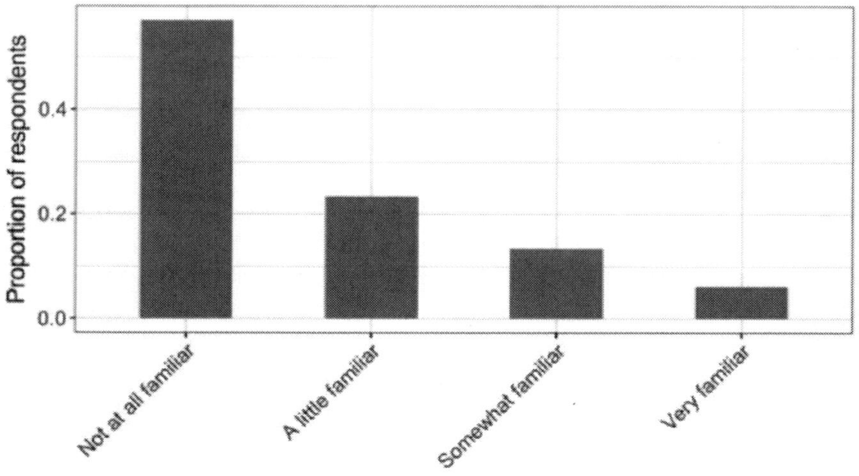

Figure 6. Familiarity with geoengineering in the U.S. [72].

Although public knowledge about solar geoengineering is low, it has been found that once the public is educated, they generally support solar geoengineering research (fig. 7). The American public tends to agree with solar geoengineering research regardless of where they fall on the political spectrum. This indicates that solar geoengineering research has the potential to receive bipartisan support. However, before bipartisan support can be garnered, the knowledge gap between solar geoengineering researchers and the general public must be abridged.

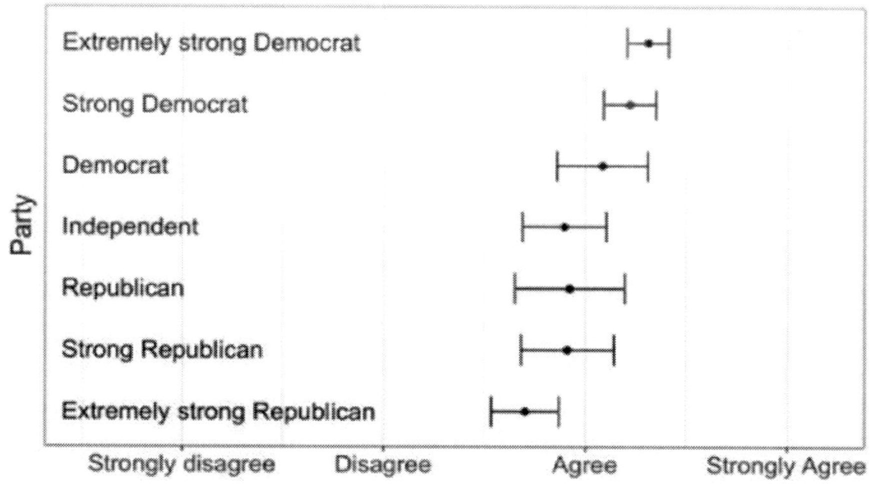

Figure 7. Support for solar geoengineering research by party [72].

Another relevant issue related to the public perception of solar geoengineering is how opinions on solar geoengineering compare to other geoengineering techniques such as CDR. A small study on New Zealanders found that CDR is generally favored over SRM geoengineering techniques [73]. This may indicate that when the public is presented with CDR, they may view solar geoengineering less favorably.

4. Policy Recommendations

4.1 Prioritizing Mitigation and Adaptation

4.1.1 The Paris Agreement

It is important to focus on emission reduction efforts in conjunction with solar geoengineering research efforts. Solar geoengineering efforts should neither overshadow nor be viewed as an alternative to emission reduction efforts. The best way to avoid this moral hazard is to commit to climate change mitigation and adaptation efforts prior to beginning solar geoengineering investigations.

Demonstrating a commitment to emission reductions to the international community prior to conducting solar geoengineering efforts is vital to avoid being accused of moral hazard. An investment in solar geoengineering research without reassuring the international community that the U.S. does not intend to implement solar geoengineering as an alternative to emission reductions could draw criticism. The UN has already established a framework via the Paris Agreement that would allow the U.S. to publicly commit to reducing emissions [23]. The Trump administration verbally withdrew from this voluntary agreement on June 1st, 2017 citing the "America First" policy as justification [74]. It is important to note that the earliest possible official withdrawal date by the United States, in accordance with Article 28 of the Paris Agreement, is not until November 4th, 2020 [23].

In 2017, President Trump stated, "I'm willing to immediately work with Democratic leaders to either negotiate our way back into Paris, under the terms that are fair to the United States and its workers" [74]. There has been no indication that such negations have been conducted yet; however, the possibility of these negotiations occurring behind closed doors cannot be ruled out. It is advised that the United States verbally rejoin the voluntary Paris Agreement to publicly demonstrate the nation's commitment to reducing greenhouse gas emissions. If necessary, the terms of the Paris Agreement should be renegotiated, but this should be done in a timely fashion. The verbal commitment to rejoin the Paris Agreement should be made as soon as possible since indicating the United State's commitment to emissions reductions is recommended prior to commencing any solar geoengineering efforts.

4.1.2 The Green Deal

In 2019, two bills were proposed in the House that outlined a general path forward to mitigate and adapt to climate change. The Green New Deal was proposed by Democratic Representative Occasio-Cortez on February 7th, 2019 as a set of goals to build climate change policy around [75]. Similarly, Republican Representative Gaetz proposed the Green Real Deal in response to the Green New Deal less than a month later April 3rd, 2019 [76]. Upon close examination of these bills, there is a significant overlap in the goals proposed by Representative Occasio-Cortez and Representative Gaetz. It is recommended that the overlap in these bills be combined to create a new bipartisan framework, tentatively dubbed "The Green Deal" (fig. 8).

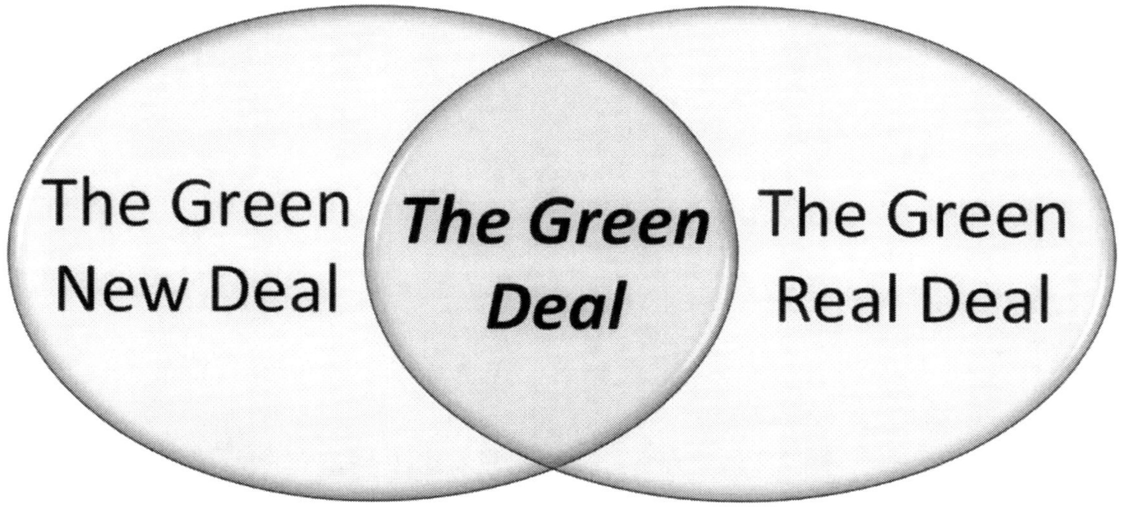

Figure 8. Coupling the "Green New Deal" with the "Green Real Deal" to establish "The Green Deal"

Although, these two documents were not identical, there is enough in common between the two to create a new cohesive set of goals. This new set of goals established by proposing "The Green Deal" could be used as an outline for a set of bipartisan policies that would ensure the prioritization of adaptation and mitigation. The goals of "The Green Deal" should include:

- Investment in clean energy infrastructure
- Economy-wide decarbonization
- Greater resiliency against climate change related disasters
- Modernization of the electric grid
- Improvements in building energy efficiency

4.2 International Oversight

4.2.1 Risk Evaluation

Since solar geoengineering is a global issue, international oversight on implementation efforts is recommended to avoid conflicts. International oversight may help mitigate the risk for the potential of a climate tug of war, rogue actors, and termination shock. It is advised that the framework previously established by the UN IPCC be used as the vehicle for international solar geoengineering oversight.

The IPCC was initially created to provide policymakers with regular scientific assessments on climate change, its implications and potential future risks, as well as to put forward adaptation and mitigation options [77]. It accomplishes this through a structure that includes four technical support units (TSU) that each have a specific set of goals (fig. 9). If the IPCC were to begin more closely examining the risks associated with solar geoengineering, at least three of these four TSU's would likely be needed.

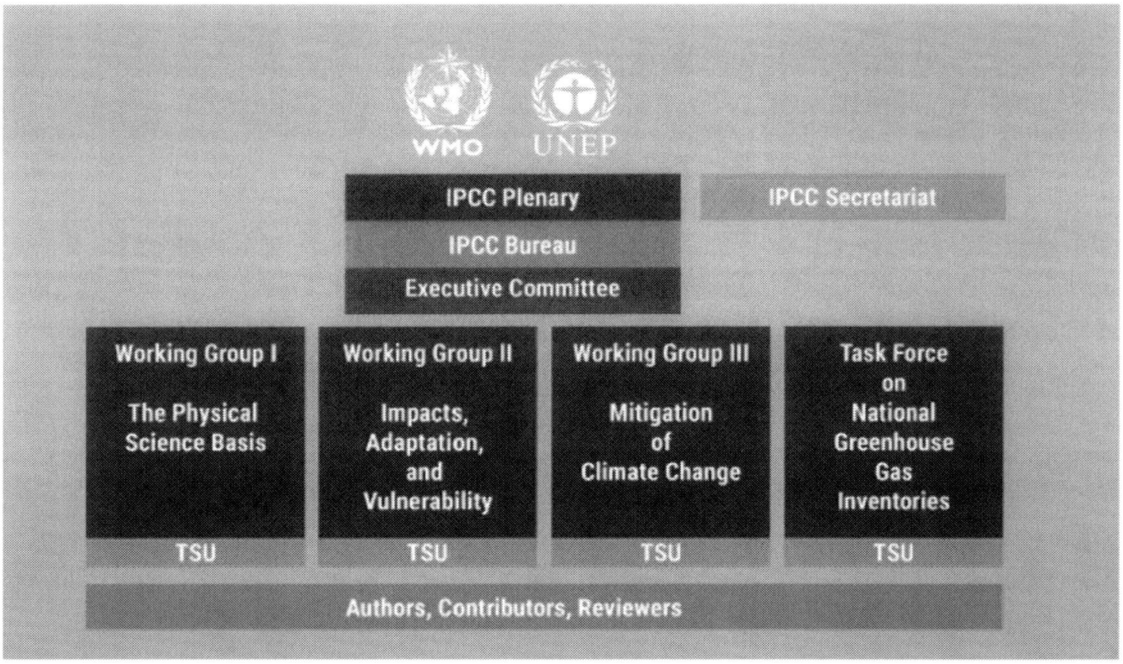

Figure 9. Structure of the IPCC [77].

Working Group I, on the physical science basis, would continue to provide information on greenhouse gases and aerosols in the atmosphere; temperature changes in the air, land and ocean; the hydrological cycle and changing precipitation (rain and snow) patterns; extreme weather; glaciers and ice sheets; oceans and sea level; biogeochemistry and the carbon cycle; and climate sensitivity [77]. This TSU already studies the effects of aerosols in the atmosphere so the investigation of the effect of stratospheric aerosol injection would be a natural extension of their current scope. Additionally, this TSU will provide valuable information for addressing the damage attribution problem previously discussed. Working Group II, on impacts, adaptation, and vulnerability assesses the global and regional effects of climate change. This assessment will provide reliable baseline data to which the effects of solar geoengineering can be compared. Working Group III, on the mitigation of climate change provides an objective evaluation of technical feasibility, cost, and enabling environments. This TSU could be expanded to assess potential solar geoengineering strategies.

One of the primary responsibilities of UN IPCC oversight on solar geoengineering should be to perform an annual risk evaluation on the emerging technology. Further evaluation of potential risks needs to be achieved prior to any consideration of deploying solar geoengineering. The environmental, health, social, and economic risks associated with solar geoengineering are still poorly understood and need much elucidation. A risk evaluation could provide more information on these risks that allow a proper cost/benefit analysis to be performed on solar geoengineering.

4.2.2 Insurance

There is a significant amount of uncertainty associated with the risk of solar geoengineering implementation even if it is implemented through a robust international framework. The establishment of international financial risk pools could help mitigate any unanticipated damages caused by solar geoengineering deployment.

Additionally, it is likely that disagreements between states may arise while discussing solar geoengineering implementation due to conflicting interests. An international risk pool could be used as a mechanism to settle disputes between nations [78]. This risk pool will also provide incentivization to limit opposition to other nations' concerns. A risk pool ensures that it would be in the international community's best interest to set the "global thermostat" to provide the most benefit to the most people. A risk pool would provide financial incentive for all participating nations to minimize solar geoengineering risks. Support for this type of insurance could also provide nations with an opportunity to demonstrate their confidence in solar geoengineering technology.

The World Bank's parametric climate insurance could serve as the model for a solar geoengineering risk pool [79]. The World Bank's index-based insurance uses objective environmental indices to protect parties from significant financial losses. Aspects of this parametric insurance should be adopted when designing a solar geoengineering risk pool. The World Bank's parametric climate insurance can be distinguished from conventional insurance by the following five key advantages [79-82]:

- It does not require causation to be demonstrated—attribution is a nonissue
- It can cover catastrophic losses
- It is oriented toward protecting against future harms (rather than resolving disputes over past damages)
- It is fundamentally contractual in that the insured agrees to pay a premium to the insurer in exchange for coverage against one or more contingencies (as opposed to the adversarial nature of legal liability)
- It is predictable

It is recommended that a solar geoengineering risk pool be established within the existing UN IPCC framework as a damage management tool. The UN IPCC Secretariat manages all funds agreed to by the IPCC, so this entity would be the most natural fit for managing the risk pool. Joining the risk pool would have to be mandatory for all participating nations for it to be effective. This could pose cost challenges to developing nations who want a say in solar geoengineering but cannot afford to participate in the risk pool. To address this challenge, the amount each nation would fund the risk pool should be proportional to their GNP. This would ensure a level playing field for all nations.

4.2.3 Deployment Monitoring and Verification

In the event that the UN IPCC were to decide to implement solar geoengineering multilaterally, a new regime will be needed to provide verification of deployment and effectiveness monitoring. This regime is needed to confirm solar geoengineering has been implemented to the degree previously agreed upon and that the deployed solar geoengineering strategy is interacting with the climate as expected. It is recommended that an international body within the UN conduct deployment monitoring and verification to ensure data reliability.

This international deployment monitoring and verification regime should be based upon the well-developed Comprehensive Nuclear-Test-Ban Treaty (CTBT) model [83]. Monitoring stations should be established across the globe and in space that will measure a variety of climatic variables. It is recommended that these stations be run by many different nations while overseen by the UN to promote robustness of collected data. The overlap between the stations owned by different nations will allow for cross-examinations of collected data to verify their accuracy.

Additionally, there are a variety of parameters that will need to be determined and agreed upon within the UN prior to solar geoengineering implementation. The following parameters will need to be confirmed prior solar geoengineering deployment (specifically stratospheric aerosol injection) to guarantee that the terms agreed upon within the UN will be properly carried out [84]:

- type of material released
- particle size
- amount released
- release altitude
- release/dispersal mechanism and area over which this is done (initial density)

4.2.4 Code of Conduct

There are a variety of ethical challenges that may be encountered while conducting solar geoengineering research. To help resolve ethical challenges, many scientists and engineers often turn to ethical codes. It is recommended that the UN IPCC establish a code of conduct for all solar geoengineering researchers to promote ethical practices. This ethical code should encourage researchers to focus on performing solar geoengineering research to benefit humanity rather than nationalistic or financial motives. It is recommended that a "Solar Geoengineering Research Code of Ethics" be established by the UN Ethics Office. This international code of conduct should be modeled after the Geoengineering Research Governance Project's "Code of Conduct for Responsible Geoengineering Research" [85].

4.3 U.S. Federal Solar Geoengineering Research Program

Internationally, China has one of the world's largest federally funded geoengineering research programs with a budget estimated to be about $3 million [86]. It is recommended that the United States establish a "U.S. Federal Solar Geoengineering Research Program" that is on par with China's research program. It is important that the United States lead the way on this innovative technology via a new research program to maintain the upper hand in international discourse. The recommended structure of the "U.S. Federal Solar Geoengineering Research Program" can be seen below (fig. 10).

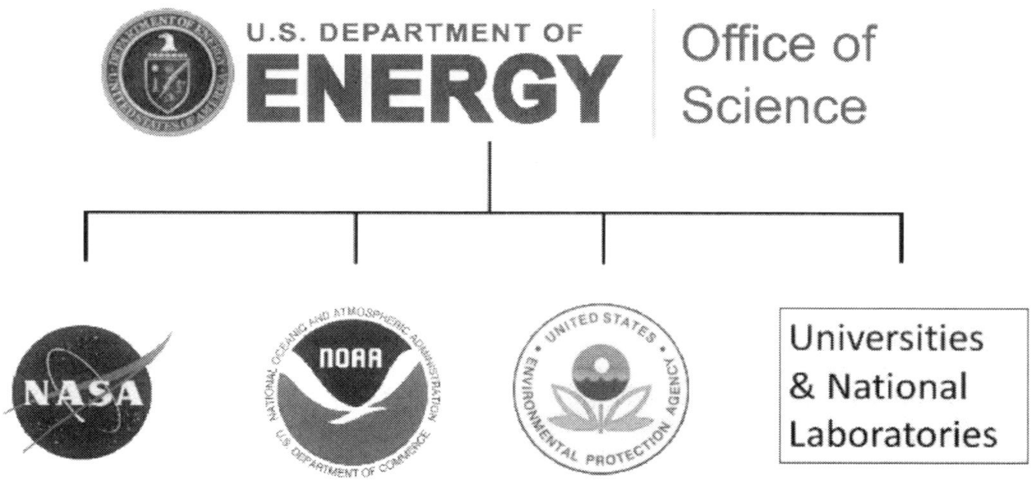

Figure 10. U.S. Federal Solar Geoengineering Program structure

It is suggested that the Department of Energy (DOE) oversee the "U.S. Federal Solar Geoengineering Program" through the Office of Science. The DOE Office of Science would distribute around $3 million in grants to national laboratories and universities to pursue a better fundamental understanding of the interaction of solar geoengineering with the climate. In addition, the grant would fund efforts to study solar geoengineering deployment mechanisms. The National Aeronautics and Space Administration (NASA) and the National Oceanic and Atmospheric Administration (NOAA) would provide their satellite resources as directed by the DOE to aid in research efforts as needed. For all real-world atmospheric experimentation that would occur outside of a laboratory setting, it would be required that the EPA perform and environmental impact assessment (EIA) prior to conducting the experiment. After performing the EIA to determine the potential health and environmental risks of the experiment, the EPA would have the discretion to either shut down or green light each real-world experiment.

4.4 Transparency

4.4.1 Public Education

To prevent public backlash, significant effort needs to be put forth to ensure solar geoengineering efforts are transparent. However, it is important to first educate and inform the public about solar geoengineering so that they possess an understanding of the technology. This type of public education would need to involve a variety of channels by which information could be consumed. These channels could be in the form of a mass media campaign, online resources provided by the U.S. government, or even a part of school curriculum. It is likely that the combination of multiple channels will be needed to reach most Americans with this important information. It is critical for researchers and politicians to engage with the American public on concerns they may have with solar geoengineering, so that research efforts are not undermined by uninformed critics.

4.4.2 Clearinghouses

One way the public could access information on solar geoengineering research efforts is through publicly accessible clearinghouses. These clearinghouses should be based upon the structure that is outlined in the Harvard Project on Climate Agreement's "Governance of the Deployment of Solar Geoengineering" [59]. It is recommended that the DOE establish such clearinghouses as part of the U.S. Federal Solar Geoengineering Research Program that would catalog all federal solar geoengineering research in the U.S. Additionally, as much privately funded research should be cataloged in the clearinghouse as possible to provide a comprehensive database of solar geoengineering information. Each study provided in the clearinghouse should have its funding publicly clarified and it is also important that the DOE summarize the state of solar geoengineering research periodically so that it is more easily digestible by the American public and politicians.

The United States' national clearinghouse should feed into an international clearinghouse overseen by the UN. The UN and the U.S. should also encourage other nations with federal solar geoengineering research programs, such as China, to follow suit. This international clearinghouse would enhance the UN's ability to properly assess the risks associated with solar geoengineering in their annual reports. Additionally, the international clearinghouse could be shared amongst nations to promote collaboration. The establishment of not only a national, but international clearinghouse is vital to preventing mistrust amongst nations and individuals.

4.5 Establishment of Legal Regime

Currently, no U.S. or international law exists to govern the implementation of solar geoengineering to mitigate climate change. As previously discussed, Russel George demonstrated the presence of this legal vacuum when he was not prosecuted for dumping 100 tons of iron sulphate into the Pacific Ocean to geoengineer the planet. This legal void should be filled both in the U.S. and internationally to promote safe, well thought out, solar geoengineering research and implementation.

It is recommended that both the U.S. and the UN outlaw individuals from implementing solar geoengineering outside of the UN IPCC's oversight. This would discourage unilateral deployment that could potentially degrade the environment, human health, and international relations. Furthermore, the UN should enact a law to prevent individual nations from implementing solar geoengineering outside of the IPCC's framework for this same reason. Finally, it suggested that the scope of the current international EIA legal rules be expanded to encompass solar geoengineering. It has been found that the current international EIA law is inadequate to grapple with the emerging technology of solar geoengineering [87]. The expansion of international EIA law could help address the environmental and social risks associated with solar geoengineering.

5. Conclusion

Whether it is decided to implement solar geoengineering or not, one must recognize that humans are already geoengineering the planet via unintentional climate change. It is not inconceivable that one day humanity may decide to modify the climate intentionally to mitigate the damages caused by greenhouse gas emission induced climate change. However, it is critical that the United States does not fall into the moral hazard of allowing future solar geoengineering efforts to undermine emission reduction efforts. Solar geoengineering should not simply be used to mask the effects of climate change given the potential for termination shock and continued ocean acidification. Solar geoengineering should instead be used as a tool to buy humanity precious time to reduce our emissions so that irreversible climate change may be avoided.

A better understanding of the underlying mechanisms of the climate and the effects solar geoengineering may have on them must be gained. A proper cost vs benefit analysis cannot be performed to determine whether solar geoengineering should be implemented without improving this understanding. A federal program designed to pursue information on the effects of solar geoengineering could achieve this desired understanding. This information should be shared with the international community in a transparent manner so that proper oversight can be conducted by the United Nations. Since the deployment of solar geoengineering will impact everybody, the United Nations must promote cooperation and provide a framework to facilitate multilateral implementation. Simultaneously, the United Nations has a responsibility to strongly discourage unilateral implementation.

The idea of modifying the atmosphere of the earth via albedo modification is no longer radical. The unprecedented rate of the release of greenhouse emissions is creating unfavorable circumstances for a growing human population. If habitability is to be ensured, the climate must be restored to be more conducive to human life. As the window closes to prevent the release of greenhouse gases from causing irreversible warming, drastic measures such as solar geoengineering may be considered. A firm understanding of this technology must be acquired within the next ten years, before climate change could become irreversible, to evaluate the risks of its implementation and a robust framework must be established for safe deployment. The actions humanity takes over the course of the next decade will determine whether solar geoengineering will be needed. Ideally, humanity will address the global climate crisis and solar geoengineering will be unnecessary; however, if humanity fails, preparations should be made to use solar geoengineering as a last resort.

References

[1] Earth System Research Laboratory, "Recent Monthly Average Mauna Loa CO_2," National Oceanic and Atmospheric Administration, May 12, 2019.

[2] Scripps Institution of Oceanography, "Carbon Dioxide at Mauna Loa Observatory reaches new milestone: Tops 400 ppm," University of California, San Diego, May 10, 2013.

[3] Intergovernmental Panel on Climate Change, "Global Warming of 1.5 °C" United Nations, Switzerland, 2018.

[4] Brown, Patrick and Caldeira, Ken. "Greater future global warming inferred from Earth's recent energy budget" *Nature*. 552. 45-53. 2017.

[5] Brysse, Keynyn; Oreskes, Naomi; O'Reilly, Jessica and Oppenheimer, Michael, "Climate change prediction: Erring on the side of least drama?" *Global Environmental Change.* 23, 327-337. 2013.

[6] "Global temperature anomalies averaged and adjusted to early industrial baseline (1881-1910)" *Climate Central.* Figure generated from data provided by NASA GISS, NOAA NCEI and ESRL. 2018.

[7] Brown, Patrick and Caldeira, Ken. "Greater future global warming inferred from Earth's recent energy budget" *Nature Publishing Group*. 552, 45-50. 2017.

[8] Rhodium Group. "Final US Emissions Estimates for 2018" May 31, 2019.

[9] Storrow, Benjamin. "2019 Power-Sector Trends Point to a Continued Rise in U.S. Emissions" *Scientific American*. June 3, 2019.

[10] U.S. Global Change Research Program, "Fourth Climate Assessment" 2017.

[11] "The National Security Implications of Climate Change: Hearings before the Committee on Intelligence" House of Representatives, 116th Cong. E750 (Testimony of Rod Schoonover). 2019.

[12] U.S. Environmental Protection Agency, "Multi-Model Framework for Quantitative Sectoral Impact Analysis" Washington, DC, 2017.

[13] United Nations Framework Convention on Climate Change, "Paris Agreement" 2015.

[14] Carnegie Climate Geoengineering Governance Initiative, "Governing Solar Radiation Modification (SRM)" Carnegie Council, 2018.

[15] Environmental Pollution Panel, "Restoring the Quality of Our Environment" United States, The White House, President's Science Advisory Committee. 1965.

[16] Rasch, Phillip; Tilmes, Simone; Turco, Richard; Robock, Alan; Oman, Luke; Chen, Chih-Chieh (Jack); Stenchikov, Georgiy; and Garcia, Rolando. "An overview of geoengineering of climate using stratospheric sulphate aerosols" *Philosophical Transactions of The Royal Society*. 366, 4007-4037. 2008.

[17] Marchetti, Cesare. "On Geoengineering and the CO_2 Problem" *Research Memoranda*. 1976.

[18] Crutzen, Paul. "Albedo Enhancement by Stratospheric Sulfur Injections: A Contribution to Resolve a Policy Dilemma?" *Climatic Change*. An editorial essay. 2006.

[19] Jha, Alok. "Obama climate adviser open to geo-engineering to tackle global warming" *The Gaurdian*. April 8, 2009.

[20] Pidgeon, Nick; Parkhill, Karen; Corner, Adam; Vaughan, Naomi. "Deliberating stratospheric aerosols for climate geoengineering and the SPICE project" *Nature Climate Change*. 3, 451-457. 2013.

[21] Zhongguo Yingdui Qihou Bianhua de Zhengce yu Xingdong – 2012 Niandu Baogao . "China's policies and operations in response to climate change" *National Development and Reform Commission (NDRC)*. 2012.

[22] National Research Council. "Climate Intervention: Reflecting Sunlight to Cool Earth" *The National Academies Press*. 2015.

[23] United Nations. "Fifth Assessment Report of the Intergovernmental Panel on Climate Change" *IPCC*. Geneva, Switzerland. 2014.

[24] Chemnick, Jean. "U.S. Blocks U.N. Resolution on Geoengineering" Scientific American. March 15, 2019.

[25] "National Academies Launching New Study on Sunlight-Reflection Research" The National Academies of Sciences, Engineering, and Medicine News. October 16, 2018. http://www.nationalacademies.org/onpinews.

[26] Keith, David. *A Case for Climate Engineering*. Massachusetts Institute of Technology Press. 2013.

[27] Vattioni, Sandro; Weisenstein, Debra; Keith, David; Feinberg, Aryeh; Peter, Thomas; Stenke, Andrea. "Exploring accumulation-mode H_2SO_4 versus SO_2 stratospheric sulfate geoengineering in a sectional aerosol–chemistry–climate model" *Atmospheric Chemistry and Physics*. 19, 4877-4897. 2019.

[28] Visioni, Daniele; Pitari, Giovanni; Aquila, Valentina; Tilmes, Simone; Cionni, Irene; Genova, Glauco; Mancini, Eva. "Sulfate geoengineering impact on methane transport and lifetime: results from the Geoengineering Model Intercomparison Project (GeoMIP)" *Atmospheric Chemistry and Physics*. 17, 11209-11226. 2017.

[29] Irvene, Peter; Kravitz, Ben; Lawrence, Mark; Muri, Helene. "An overview of the Earth system science of solar geoengineering" *Wiley Interdisciplinary Reviews: Climate Change*. 7, 815-833. 2016.

[30] Lo, Y. T. E.; Charlton-Perez, Andrew; Highwood, Eleanor; Lott, Fraser. "Best Scale for Detecting the Effects of Stratospheric Sulfate Aerosol Geoengineering on Surface Temperature" *Earth's Future*. 12, 1660-1671. 2018.

[31] Robock, Alan; Oman, Luke; Stenchikov, Georgiy. "Regional climate responses to geoengineering with tropical and Arctic SO_2 injections" *Journal of Geophysical Research Atmospheres*. 16, 1-15. 2008.

[32] Keutsch Research Group. "SCoPEx" *Harvard University*. https://projects.iq.harvard.edu/keutschgroup/scopex.

[33] Latham, John; Gadian, Alan; Fournier, Jim; Parkes, Ben; Wadhams, Peter; Chen, Jack. "Marine cloud brightening: regional applications" *Philosophical Transaction of The Royal Society*. 2014.

[34] Fleming, James. *Fixing The Sky: The Checkered History of Weather and Climate Control*. Columbia University Press. 2012.

[35] Lee, Joonsuk; Yang, Ping; Dessler, Andrew; Gao, Bo-Cai; Plantick, Steven. "Distribution and radiative forcing of tropical thin cirrus clouds" *Journal of the Atmospheric Sciences*. 12, 3721-3731. 2009.

[36] Storelvmo, T.; Kristjansson, J. E.; Muri, H.; Pfeffer, M.; Barahona, D.; Nenes, A. "Cirrus cloud seeding has potential to cool climate" *Geophysical Research Letters*. 40, 178–182. 2013.

[37] Marshall, Tom. "Reflective crops could soften climate change blow" *NERC Planet Earth*. January 20, 2009.

[38] Lee, Xuhui et al. "Observed increase in local cooling effect of deforestation at higher latitudes" *Nature Letters* 479, 384-387. 2011.

[39] Akbari, Hashem; Mathews, H; Seto, Donny. "The long-term effect of increasing the albedo of urban areas" *Environmental Research Letters*. 7, 1-10. 2012.

[40] Field, L et al. "Increasing Arctic Sea Ice Albedo Using Localized Reversible Geoengineering" *Earth's Future*. 882-901. 2018.

[41] "Science benefits from diversity" *Nature*. 5. June 6, 2018.

[42] Nowack, Peer; Abraham, Nathan; Braesicke, Peter; and Pyle, John. "Stratospheric ozone changes under solar geoengineering: implications for UV exposure and air quality" *Atmospheric Chemistry and Physics*. 16, 4191-4203. 2016.

[43] Muri, Helene; Niemeier, Ulrike; and Kristjánsson, Jón. "Tropical rainforest response to marine sky brightening climate engineering" *Geophysical Research Letters*. 2951-2960. 2015.

[44] Ferraro, A.; Charlton-Perez, A.J.; and Highwood, E.J. "Stratospheric dynamics and midlatitude jets under geoengineering with space mirrors and sulfate and titania aerosols" *Journal of Geophysical Research: Atmospheres*. 414-429. 2104.

[45] "You can blame climate change for the cold weather" *CBC Kids News*. February 1, 2019.

[46] Lauvset, Siv; Tjiputra, Jerry; and Muri, Helene. "Climate engineering and the ocean: effects on biogeochemistry and primary production" *Biogeosciences*. 14, 5675-5691. 2017.

[47] Xia, Lili; Nowack, Peer; Tilmes, Somine; and Robock, Alan. "Impacts of stratospheric sulfate geoengineering on tropospheric ozone" *Atmospheric Chemistry and Physics*. 17, 11913-11928, 2017.

[48] Environmental Protection Agency. "Understanding Global Warming Potentials" https://www.epa.gov/ghgemissions/understanding-global-warming-potentials

[49] Ji, Duoying; Fang, Songsong; Curry, Charles; Kashimura, Hiroki; Watanabe, Shingo; Cole, Jason; Lenton, Andrew; Muri, Helene; Kravitz, Ben; and Moore, John. "Extreme temperature and precipitation response to solar dimming and stratospheric aerosol geoengineering" *Atmospheric Chemistry and Physics.* 18, 10133-10156. 2018.

[50] Bal, Prasanta; Pathak, Raju; Mishra, Saroj; and Shany, Sandeep. "Effects of global warming and solar geoengineering on precipitation seasonality" *Environmental Research Letters*. 14. 2019.

[51] Tollefson, Jeff. "Iron-dumping ocean experiment sparks controversy" *Nature News*. May 23, 2017.

[52] Fountain, Henry. "A rogue climate experiment outrages scientists" *The New York Times*. October 28, 2012.

[53] Parson, Ted. "Canada's ocean fertilization flap, and its significance" *Legal Planet*. October 18, 2012.

[54] Jones, A.; Haywood, J. M.; Alterskjaer, K.; Boucher, O.; Cole, J. N. S.; and Curry, C. L. "The impact of abrupt suspension of solar radiation management (termination effect) in experiment G2 of the Geoengineering Model Intercomparison Project (GeoMIP)" *Journal of Geophysical Research: Atmospheres*. 118, 9743–9752. 2013.

[55] Matthews, H.D. and Caldeira, K. "Transient climate-carbon simulations of planetary geoengineering" *Proc Natl Acad Sci*. 104, 9949–9954. 2007.

[56] Sugiyama, Masahiro; Arino, Yosuke; Kosgui, Takanobu; Kurosawa, Atsushi; and Watanabe, Shingo. "Next steps in geoengineering scenario research: limited deployment scenarios and beyond" *Climate Policy*. 18, 681-689. 2017.

[57] Svoboda, Toby; Irvine, Peter; Callies, Daniel; and Sugiyama, Masahiro. "The potential for climate engineering with stratospheric sulfate aerosol injections to reduce climate injustice" *Journal of Global Ethics*. 14, 353-368. 2018.

[58] Trisos, Christopher; Amatulli, Giuseppe; Gurevitch, Jessica; Robock, Alan; Xia, Lili; and Zambri, Brian. "Potentially dangerous consequences for biodiversity of solar geoengineering implementation and termination" *Nature Ecology and Evolution*. 3, 475-482. 2018.

[59] Harvard Project on Climate Agreements. "Governance of the deployment of solar geoengineering Harvard Project on Climate Agreements" February 2019.

[60] Bas, M. and Mahajan, A. "Contesting the Climate: Security Implications of Geoengineering and Counter-geoengineering." September 28, 2018.

[61] Brutschin, Elina and Schubert, Samuel. "Icy waters, hot tempers, and high stakes: Geopolitics and Geoeconomics of the Arctic" *Energy Research & Social Science*. 16, 147-159. 2016.

[62] Stringer, Lindsay; Dyer, Jen; Reed, Mark; Dougill, Andrew; Twyman, Chasca; and Mkwambisi, David. "Adaptations to climate change, drought and desertification: local insights to enhance policy in southern Africa" *Environmental Science and Policy*. 12, 748-765. 2009.

[63] United Nations. "Convention on the prohibition of military or any other hostile use of environmental modification techniques" New York. December 10, 1976.

[64] Lo, Y.T.; Charlton-Perez, Andrew; Lott, Fraser; and Highwood, Eleanor. "Detecting sulphate aerosol geoengineering with different methods" *Nature Scientific Reports*. 6, 1-10. 2016.

[65] Sobel RS, Nesbit TM. "Automobile safety regulation and the incentive to drive recklessly: evidence from NASCAR" *South. Econ. J.* 74, 71–84. 2007.

[67] Peltzman S. "The effects of automobile safety regulation" *J. Polit. Econ*. 83, 677–726. 1975.

[68] Cassell MM, Halperin DT, Shelton JD, Stanton D. "Risk compensation: the Achilles' heel of innovations in HIV prevention?" *Br.Med.J* 332, 605–607. 2006.

[69] Sadowski NC, Sutter D. "Hurricane fatalities and hurricane damages: are safer hurricanes more damaging?" *South. Econ. J.* 72, 422–432. 2005.

[70] Clark JR, Lee DR. "Too safe to be safe: some implications of short- and long-run rescue Laffer curves" *East. Econ. J.* 23, 127–137. 1997.

[71] Morrow, David. "Ethical aspects of the mitigation obstruction argument against climate engineering research" *Philosophical Transactions of the Royal Society A: Mathematical, Physical and Engineering Sciences.* 372, 1-14. 2014.

[72] Mahajan, Aseem; Tingley, Dustin; and Wagner, Gernot. "Fast, cheap, and imperfect? US public opinion about solar geoengineering" *Environmental Politics.* 3, 523-543. 2019.

[73] Wright, Malcom; Teagle, Damon; Feetham, Pamela. "A quantitative evaluation of the public response to climate engineering" *Nature Climate Change*. 2, 106-110. 2014.

[74] Trump, Donald. "Statement by President Trump on the Paris Climate Accord" *The white House*, The United States Government, June 1, 2017. https://www.whitehouse.gov/briefings-statements/statement-president-trump-paris-climate-accord/

[75] The Green New Deal. H. Res. 109., 116th Congress. 2019.

[76] The Green Real Deal. H. Res. 288., 116th Congress. 2019.

[77] The Intergovernmental Panel on Climate Change. "About the IPCC" 2019. https://www.ipcc.ch/about/.

[78] Horton, Joshua and Keith, David. "Multilateral parametric climate risk insurance: a tool to facilitate agreement about deployment of solar geoengineering?" *Climate Policy*. 1-7. 2019.

[79] World Bank. "Sovereign climate and disaster risk pooling: World Bank technical contribution to the G20" Washington, D.C. 2017.

[80] Horton, J. B. "Parametric insurance as an alternative to liability for compensating climate harms" *Carbon and Climate Law Review*. 12, 285–296. 2018.

[81] Mechler, R., Bouwer, L. M., Schinko, T., Surminski, S., & Linnerooth-Bayer, J. *Loss and damage from climate change: Concepts, methods and policy options*. Cham, Switzerland: Springer. 2019.

[82] Surminski, S., Bouwer, L. M., & Linnerooth-Bayer, J. "How insurance can support climate resilience" *Nature Climate Change*. 6, 333–334. 2016.

[83] Preparatory Commission for the CTBTO (Comprehensive Test Ban Treaty Organization). Verification Regime. 2018. www.ctbto.org/verification-regime.

[84] Smith P., L. Wickman, I. Min, and S. Beck. "Feasibility of Space-Based Monitoring for Governance of Solar Radiation Management Activities." AIAA SPACE 2010 Conference & Exposition. 8767. 2010.

[85] Hubert, Anna-Maria. "Code of conduct for Responsible Geoengineering Research" Geoengineering Research Governance Project. October 2017.

[86] Edney, Kingsley and Symons, Jonathon. "China and the blunt temptations of geo-engineering: the role of solar radiation management in China's strategic response to climate change" *The Pacific Review*. 3, 307-332. 2014.

[87] Craik, Neil. "International EIA Law and Geoengineering: Do Emerging Technologies Require Special Rules?" *Climate Law*. 5, 111-141. 2015.

582346 Addressing Management Challenges Within the Nuclear Weapons Complex to Streamline Plutonium Disposition

The U.S. lacks a nuclear waste disposition program. Despite demonstrating technical competency, the country has yet to implement a policy, notably for its excess defense plutonium. The Department of Energy (DOE), the agency responsible for management of supporting activities for the nuclear weapons complex, historically has demonstrated management challenges. Unless the agency can overcome the challenges that exist in its current state, the U.S. will be unable to move forward with implementing any disposition policy. The DOE's management practices in offices overseeing nuclear waste disposition and lack of action has generated public mistrust. In order to effectively implement disposition policy, in the future, the DOE must address these challenges and rebuild its relationship with the general public.

580336 A Chemical Engineer's Overview of Artificial Intelligence and Machine Learning

This hands-on workshop will start with a high level overview of Artificial Intelligence and Machine Learning, citing case studies of their development, and guide participants through identifying where these concepts can be applied in their chemical engineering practice.

How Virtual Advisors Create Practical Employee Efficiency Boosts in the Chemical Engineering Workplace

Alec Walker, DelfinSia, Inc, Houston, TX, USA.

Abstract

In an effort to explore the ability of text analytics software to augment human performance within the chemical engineering domain, a virtual advisor designed for the domain was used by chemical engineering practitioners in a controlled experiment. The experiment focused on the common industry problem of inefficient and incomplete extraction of relevant knowledge and data from within the organization's digital unstructured textual (DUT) files in order to resolve issues. The experiment focused on hydrocarbon refinery content within the broader chemical engineering domain. The virtual advisor used in the experiment employs machine learning and natural language processing to interpret the intent of human queries and instantly return knowledge and data from across multiple DUT files that is relevant to the queries. The advisor has shown capacity in commercial applications within the industry, but the key details of those results remain proprietary to the corporate clients.

While this study cannot name the individual participants in the experiment, the corporations they represent, or their proprietary issues explored using the virtual advisor during the experiment, the experimental results comparing the performance of the individuals with and without the virtual advisor are shared in detail. The results suggest that the virtual advisor increases the speed and accuracy of issue resolution in the chemical engineering domain over typical incumbent tools in a base case control scenario by significant margins. This in turn suggests that chemical engineering is an appropriate domain for the productive application of virtual advisors based on "artificial intelligence" methods including machine learning and natural language processing. The implications of the results are explored, and additional larger and more open studies in this area are recommended.

Introduction

The process-driven nature of the chemical engineering domain means that knowledge and data potentially useful in resolving issues already resides in the organization's internal documentation. When issues arise that demand resolutions, practitioners first use incumbent methods to extract relevant information from these documents before then applying their skills putting it to use. One incumbent method is to store files in hard copy form in a physical library, creating logistical challenges and risks. Leading organizations have digitized their files on shared folders. They often use a combination of inter-file methods, intra-file methods, and human reading to find data and knowledge buried in the organization:

1) The most common inter-file method is keyword-based enterprise search. This type of search tool relies on keywords to find a subset of the total files in the organization that contain the keywords entered by the practitioner into a search field. The practitioner then browses through the files in the search results, opening whichever seem relevant individually.
2) The most common intra-file method is keyword-based textual search. This functionality tends to be built into the software handing the individual file. When a potentially useful document found from inter-file search is opened, the practitioner typically has the option to use "Ctrl + F" to type in the key words again and browse through each of their occurrences within the document one at a time.
3) The final step in the incumbent process is simple human reading. Following the intra-file keyword search, if the document seems relevant, the practitioner reads it carefully looking

for useful knowledge and data. If the document does not seem relevant, the practitioner closes it and continues to browse the search results from the enterprise search.

Processes like these account for 80% of knowledge worker time in the energy industry (1), and that percentage is likely at least as high across the entire chemical engineering domain. Even though the relevant knowledge and data are contained in the organization's files, and despite the fact that the job of the knowledge workers is to put this information to use, locating and extracting the information still accounts for the vast majority of professional time (2).

Recent advances in the field of natural language processing now allow computers to perform much more advanced search than that based simply on keywords. The cutting edge in natural language processing, however, still requires domain-specific applications of the technology in order to be effective. As a result, domains are each beginning to separately experiment with natural language processing to explore its potential (3). With the goal of exploring the potential of natural language processing in the chemical engineering domain, a virtual advisor based on natural language processing was evaluated against an expedited incumbent method as described above. The virtual advisor takes in domain-specific questions rather than key words. The goals of the advisor are to reach faster and more accurate resolutions to the common issues requiring reference to digital unstructured textual (DUT) files. The virtual advisor attempts to accomplish these goals through a single search across and within multiple DUT files and by comparing each option within the digital text to the intent of the question asked.

Another recent study with a team of engineering consultants at a major multinational chemical company showed that this virtual advisor was able to successfully mine DUT files to extract information for practitioners more efficiency and effectively than their incumbent solutions. According to one engineer involved, "[The tool] helps me solve the problem of being able to search for information within reports which I already know contain that information but can't remember which report or where in the report." This study follows that industry trial for the purpose of creating detailed and accurate data open to the community pertaining to the effectiveness of the virtual advisor as a proxy for the potential of text analytics in the chemical engineering domain generally.

Key Terms

"Artificial Intelligence" – A popular term describing techniques in data science and computer science.

Machine Learning – A method in the field of computer science involving algorithms that iteratively adjust models to fit data.

Natural Language – What scientists call the form of language humans use to interact with each other as opposed to what humans use in traditional interactions with machines. Natural language does not imply English language.

Natural Language Processing – The science of using computers to process natural language. This could involve using algorithms to process audio files to convert speech into text, or it could involve using algorithms to process video files to read the movement of lips. This paper focuses on text analytics.

Text Analytics – The branch of natural language processing dealing with digital text written in natural language.

Digital Unstructured Textual (DUT) Files – These are files on computers that humans use to save and reference information. They are almost always written by a human, and they are composed of textual information. Typically, these files are saved on a hard drive or a shared folder. They might be reports, emails, memos, compliance documents, manuals, job descriptions, etc. They might end in such extensions as .ppt, .doc, .txt, .pdf, etc. This paper, when opened on a computer, is a good example of a DUT file. Note that some DUT files are mere images of typed text as opposed to actual typed text, adding one more layer of complexity to the process of employing text analytics. DUT files make up 90% of the total information across all organizations.

Methodology

The methodology involved a focus group experiment. Research and design chemical engineers from the hydrocarbon refining industry proposed resolutions to common workplace issues through reference to a set of DUT files in two scenarios: a base case using expedited incumbent reference tools and a text analytics case using a custom-built virtual advisor. The speed with which results were obtained was measured across both scenarios quantitatively (rounded to the nearest half minute), and the effectiveness of the solutions obtained was determined qualitatively from surveying the chemical engineers following the experiment.

The corpus of DUT files used in the experiment came from an open online forum for hydrocarbon refinery operations engineers started in 2009. The forum contains more than 6,000 DUT files in the form of typed discussion threads. In the base case, expedited incumbent reference tools for the experiment included a standard keyword search option and a browsable window showing all content from multiple DUT files. The standard keyword search option enables entering a keyword to have the system highlight that key word the entire body of DUT content. This effectively already cuts out the inter-file search step, giving the base case scenario an advantage over the typical scenario within a chemical engineering organization. In the text analytics case, the virtual advisor served as the only reference tool. The virtual advisor uses text analytics to interpret the intent of questions typed by the chemical engineer. The virtual advisor then processes the entire body of content within the DUT files and returns relevant passages to the chemical engineer all at once in the same window.

Seven chemical engineers with a combined total of 29 years of professional experience were involved in the focus group experiment. Four had backgrounds primarily in process improvement research, and three had backgrounds primarily in reactor design. Each were asked individually to consider their professional experience and come up with four issues they have faced which required reference to DUT files in order to be resolved. These issues were summarized generally, without specifics of the companies, reactors, or other details. For example, "The amine unit began producing problematic foam", and "The ratio of the height to the number of trays needed to be optimized". Despite anonymization of the identities of the chemical engineers involved with the study, the individual issues were deemed proprietary and are unable to be shared in this paper.

The chemical engineers were then sorted into groups based on their stated lack of professional experience with each of the issues. Seventeen of the issues had not been experienced by at least three of the seven chemical engineers. For a randomly selected nine of these seventeen issues, the members of the corresponding group lacking related experience underwent a trial using the base case scenario to craft a resolution for the issue. For the other eight of these seventeen issues, the members of the corresponding group lacking related experience underwent a trial using the text analytics scenario to come up with a resolution for the issue. We called these people "Proposers". For each trial, a time limit of ten minutes was imposed.

The proposed resolutions to the issues were then scored on their relevance and accuracy by the chemical engineers who mentioned having prior experience with these issues. We called these people "Evaluators". Comparisons between the scores obtained from each of the two scenarios were then made and explored. Concerning relevance of a proposal, evaluators were asked, "Is the proposal related to the issue?". Concerning accuracy of a proposal, evaluators were asked, "Could the proposal potentially be useful in resolving the issue?". The evaluators used the scoring system in Table 1 to respond to these questions in the case of each issue + proposal pair.

Table 1. Scoring System for Proposed Resolutions from Experiment Scenarios

Score	0	1	2	3
Relevance	No Relevancy	Little Relevancy	Mostly Relevant	Entirely Relevant
Accuracy	No Accuracy	Little Accuracy	Mostly Accurate	Entirely Accurate

Notes:

- Relevancy and accuracy scores of trials in which no proposals were made were unassigned (marked as "NA") rather than scored zero.
- Any proposal with a score of at least one was considered "relevant" in the resulting tables.
- Accuracy scores of proposals with a relevancy score of zero were unassigned (marked at "NA") rather than scored zero.
- Average percentages for relevance and accuracy were based on the range of options within Table 1. For example, an average accuracy score of zero was deemed 0% accurate, an average accuracy score of one was deemed 33% accurate, an average accuracy score of two was deemed 66% accurate, and an average accuracy score of three was deemed 100% accurate.

Results and Discussion

The results for the base case scenario show a relatively lower proposal rate, average relevancy, and average accuracy, as well as a relatively high average amount of time to research and craft a proposal.

Table 2. Scoring Summary for Base Case Scenario

Issue #	Proposers	Proposals	Avg. Time (Min)	Evaluators	Avg. Relevancy	Avg. % Relevancy	Relevant Proposals	Avg. Accuracy	Avg. % Accuracy
1	3	3	9.5	4	0.33	11%	1	2	67%
2	4	3	10	3	1.33	44%	3	1	33%
3	4	0	10	3	NA	NA	0	NA	NA
4	4	1	9.5	3	0	0%	0	NA	NA
5	5	5	7.5	2	2.2	73%	5	1.6	53%
6	3	3	10	4	1	33%	1	1	33%
7	3	3	2	4	3	100%	3	2.33	78%
8	3	0	10	4	NA	NA	0	NA	NA
9	3	3	8	4	1	33%	2	1	33%

The results for the text analytics scenario show a relatively high proposal rate, average relevancy, and average accuracy, as well as a relatively low average amount of time to research and craft a proposal.

Table 3. Scoring Summary for Text Analytics Scenario

Issue #	Proposers	Proposals	Avg. Time (Min)	Evaluators	Avg. Relevancy	Avg. % Relevancy	Relevant Proposals	Avg. Accuracy	Avg. % Accuracy
10	3	3	4	4	2	67%	3	1.67	56%
11	4	4	2	3	2.75	92%	4	2.5	83%
12	4	4	3	3	2	67%	4	2.75	92%
13	3	2	7	4	3	100%	2	1.5	50%
14	4	4	4	3	3	100%	4	1.25	42%
15	3	3	4	4	2.33	78%	3	1.33	44%
16	4	4	2	3	3	100%	4	2.5	83%
17	3	1	10	4	0	0%	0	NA	NA

In the base case scenario, there were 32 total opportunities for proposals, but there were only 21 proposals (66% of total opportunities) made by the end of the ten minutes in each trial. In the text analytics scenario, there were 28 total opportunities for proposals, with 25 proposals (89% of total opportunities) made by the end of the ten minutes in each trial. There are two possibilities for each case of these discrepancies: either the time limit was too short for the chemical engineer to find relevant information to formulate a proposal, or the DUT files lacked relevant information. While lack of relevant information was reportedly a factor, twelve proposers reported running out of time and having to submit no proposal or half-formulated proposals in the base case scenario, whereas only two proposers reported running out of time in the text analytics scenario. The text analytics scenario also shows a smaller average time requirement for proposal crafting (4.5 minutes across all proposal opportunities for the scenario) than the base case scenario (8.5 minutes across all proposal opportunities for the scenario). These results strongly suggest that the virtual advisor was able to drastically speed up the process of crafting resolutions to common issues. One participant explained, "I just typed the issue into the virtual advisor interface and then instantly found all the relevant information from the files. The advisor put all the answers from the files on the screen for me so I could just read them and combine them into my answer each time...The keyword search makes me scroll through each result one at a time, and most are not actually relevant to the issue at hand. It's much slower."

These results are important to consider in the context of a typical working environment. The base case scenario treated the body of DUT files as the equivalent of a single DUT file, cutting out the inter-file search and instead allowing the chemical engineer to jump to instances of keyword matches one at a time instantly. Despite this advantage, and even though all trials were capped at ten minutes, the average trial in the base case scenario still took 189% as long as the average trial in the text analytics scenario.

The average relevancy score of the finished proposals in the base case scenario was 42%, whereas the average relevancy score of the finished proposals in the text analytics scenario was 75%. This difference was likely attributable to the ability of the virtual advisor to filter out irrelevant information. As the experiment was controlled to ensure the chemical engineers were facing issues outside their previous experiences, there was a possibility that participants could make an error in judgment thinking that certain information was relevant when it actually was not. In addition, some issues appear to be more challenging than do others, as evidenced by longer trial times and fewer proposals from proposers. The disproportionate number of outliers in the base case scenario suggests that more is going on than challenging questions. In the base case scenario, there were 21 total proposals made, but there were only 15 relevant proposals (71% of proposals made and 47% of total opportunities). In the text analytics scenario, there were 25 total proposals made, with 24 of them relevant (96% of proposals made and 86% of total opportunities). It is speculated that the natural language processing of the virtual advisor assisted in ensuring relevancy. One participant said, "With [the virtual advisor], when you can see the answers together, you can mentally group like with like. You know what to disregard, so you don't go down the wrong track."

In the base case scenario, the average accuracy score of relevant proposals was 50%. In the text analytics scenario, the average accuracy score of relevant proposals was 64%. Despite having four more total proposal opportunities than the text analytics scenario, the base case scenario yielded a much smaller sample size of relevant proposals, as 53% (17 out of 32) total proposal opportunities in the base case scenario did not convert to relevant proposals. A previous study involving the same virtual advisor found that subject matter experts considered the results of the advisor to be similar to the results returned by a human with experience in the domain (4). Like the higher percent relevance of proposals, the higher accuracy of relevant results in the text analytics scenario is also speculated to be the result of the natural language processing employed by this virtual advisor.

Conclusions

Notable in the results is the distinction between the augmentation of human performance and the replacement of humans. Intuitively, it makes sense that trained machines outperform humans in finding data across enormous and growing bodies of DUT files. Intuitive, it also makes sense that trained humans outperform machines at interpreting relevant information presented by trained machines. Text analytics appears to show strong potential as a valuable tool when applied to the chemical engineering domain. The virtual advisor in question emerges as apparently faster and more accurate than traditional information extraction methods. This appears to be a function of the effectiveness of the virtual advisor as well as a function of the ineffectiveness of incumbent solutions. This conclusion fits with other findings showing that human performance can be augmented in the industry by advanced data processing techniques (5). Given the improvements in efficiency and performance, the virtual advisor (and text analytics methods more generally) could generate enormous value in the chemical engineering domain in areas including increased safety, improved production, and reduced costs.

Additional testing is advisable. Most importantly, the focus group experiment could benefit from increased transparency on the questions asked and the participating individuals and companies. The focus group experiment could also benefit from a more tightly controlled corpus of data to ensure that accurate resolutions to the trial issues are contained within. Likewise, the experiment could benefit from a more standardized set of trial issues with measured attempts to control equality in their difficulty and commonality. A much larger set of proposers and trial issues and a more expert and assuredly impartial set of evaluators would further help to firm the conclusions made in this report. Furthermore, while the trials in the experiment were sourced from participants, and while each participant uniformly agreed that the trials resembled those commonplace within their organizations, more data confirming the specifics of why and how need to be collected from the organizations and any appropriate changes to the focus group setting duly incorporated. Finally, the study needs to be expanded to include other areas of the chemical engineering domain.

General research into the nature of what makes text analytics successful within a domain and comparative controlled evaluation of text analytics across multiple domains will complement focus group studies such as this. Larger scale studies likely require more concerted attention from across major professional societies linked to the industry.

References

1. Larsen, Åshild Hanne, Equinor CIO Keynote Address, *Society of Petroleum Engineers*, SPE Digital Transformation Study Group, Digital Transformation Annual Congress 2019, May 10, Houston, Texas, USA.
2. Feblowitz, Jill (2013), “Analytics in Oil and Gas: The Big Deal about Big Data,” *Society of Petroleum Engineers*, SPE Digital Energy Conference, 5-7 March, The Woodlands, Texas, USA, SPE-163717-MS.
3. Jensen, Jesse W. and Perrons, Robert K. (2015), “Data as an asset: What the oil and gas industry can learn from other sectors about ‘Big Data’,” *Energy Policy*, Volume 81, June 2015, pp. 117-121.
4. Walker, Alec (2019), “Streamline digital information searches with advanced data techniques,” *Hydrocarbon Processing*, January, pp. 23-24.
5. Sumbal, M., Tsui, E. and See-to, E. (2017), "Interrelationship between big data and knowledge management: an exploratory study in the oil and gas sector", Journal of Knowledge Management, Vol. 21 No. 1, pp. 180-196.

580204 Global Teams: Technology & People

Hear some of the practical aspects of working with teams spread across the world: time zones and cultures. This includes both cost effective technology solutions and the critical aspects of building and managing teams you do not see on a daily or even regular basis. Tips, tricks and key success factors will be presented and discussed.

575893 Triplet-Triplet Annihilation Upconversion-Based Nanosensors for Fluorescent Detection of Potassium

Ionophore-based optical nanosensors offer many advantages over traditional organic indicators for biological monitoring including response tuning, multi-analyte detection, and ease of development. Typical ionophore-based nanosensors use nile blue-derived indicators called chromoionophores which must contend with strong background absorption, autofluorescence, and scattering in biological samples that limit their usefulness. Here, we demonstrate potassium-selective nanosensors that utilize triplet-triplet annihilation upconversion to minimize potential optical interference in biological media and a pH-sensitive quencher molecule to modulate upconversion intensity in response to changes in analyte concentration. A triplet-triplet annihilation dye pair (platinum (II) octaethylporphyrin and 9,10-diphenylanthracene) was integrated into nanosensors containing an analyte binding ligand (ionophore), charge-balancing additive, and a pH indicator quencher.

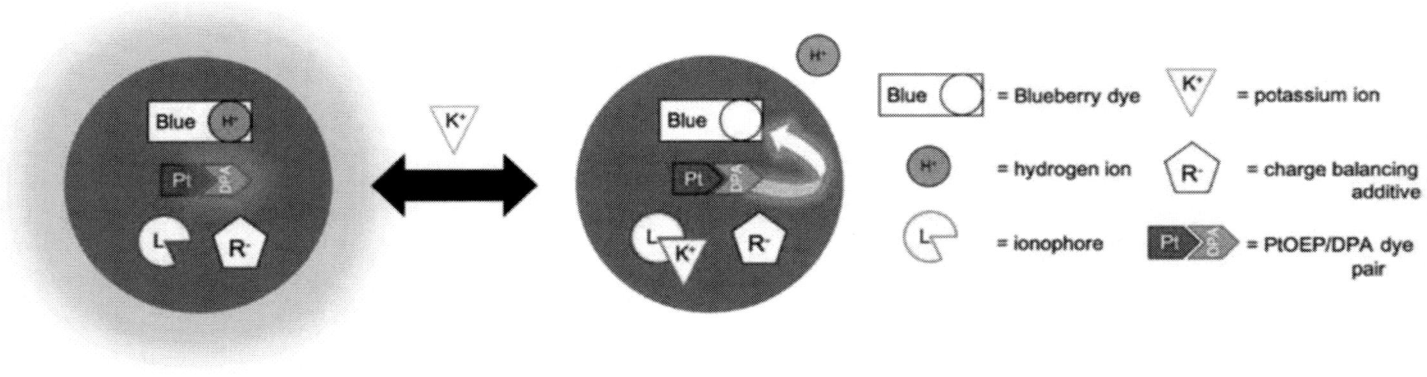

Figure 1. Schematic of nanosensor components co-loaded into the nanosensor matrix and fluorescence quenching mechanism in response to decreasing potassium (K^+) concentrations.

In comparison to lanthanide systems, TTA-UC requires relatively low power excitation and can be wavelength tuned by choice of sensitizer and annihilator. The sensitizer-annihilator dye pairs are organic dyes that are often commercially available, soluble in the organic phase, and able to be implemented into IBOS. TTA-UC has been utilized in several forms as a biological imaging agent. TTA-UC has found limited application in sensing platforms. Thus far, it has been limited to Borisov et al. creating a polymer optode film to directly sense oxygen. However, there are already numerous optical oxygen sensing platforms available. To our knowledge, there have been no reports of TTA-UC used in a more general sensing platform such as those based on ionophores. Nanosensor response to potassium was shown to be reversible and stable for three days. In addition, the nanosensors are selective against sodium and calcium (selectivity coefficient of -0.23 for calcium), the two main interfering ions in biological samples.

578077 Glucose-Responsive Nanoparticles Based on Enzymatic Sensors for Self-Regulated Insulin Delivery

Glucose-responsive nanoparticles based on enzymatic sensors

for self-regulated insulin delivery

Lisa R Volpatti, Morgan A. Matranga, Abel B. Cortinas, Robert Langer, Daniel G. Anderson

Koch Institute & Department of Chemical Engineering, MIT, Cambridge, MA 02139

Motivation: Diabetes mellitus is a disease characterized by poor glycemic control which often leads to severe complications including cardiovascular disease and kidney failure. Many diabetic patients continually monitor their blood sugar and self-administer multiple daily doses of insulin to combat hyperglycemia. Most commercial glucometers rely on the enzymatic sensor, glucose oxidase (GOx), to detect glucose levels. To better mimic native insulin activity, the GOx sensor can also be incorporated into an insulin delivery system. By coupling the sensing with a responsive polymeric material, insulin can be released in a self-regulated manner. Here, we report a glucose-responsive insulin delivery system based on the enzymatic sensing of glucose oxidase that achieves superior glycemic control compared to systems lacking the glucose sensor.

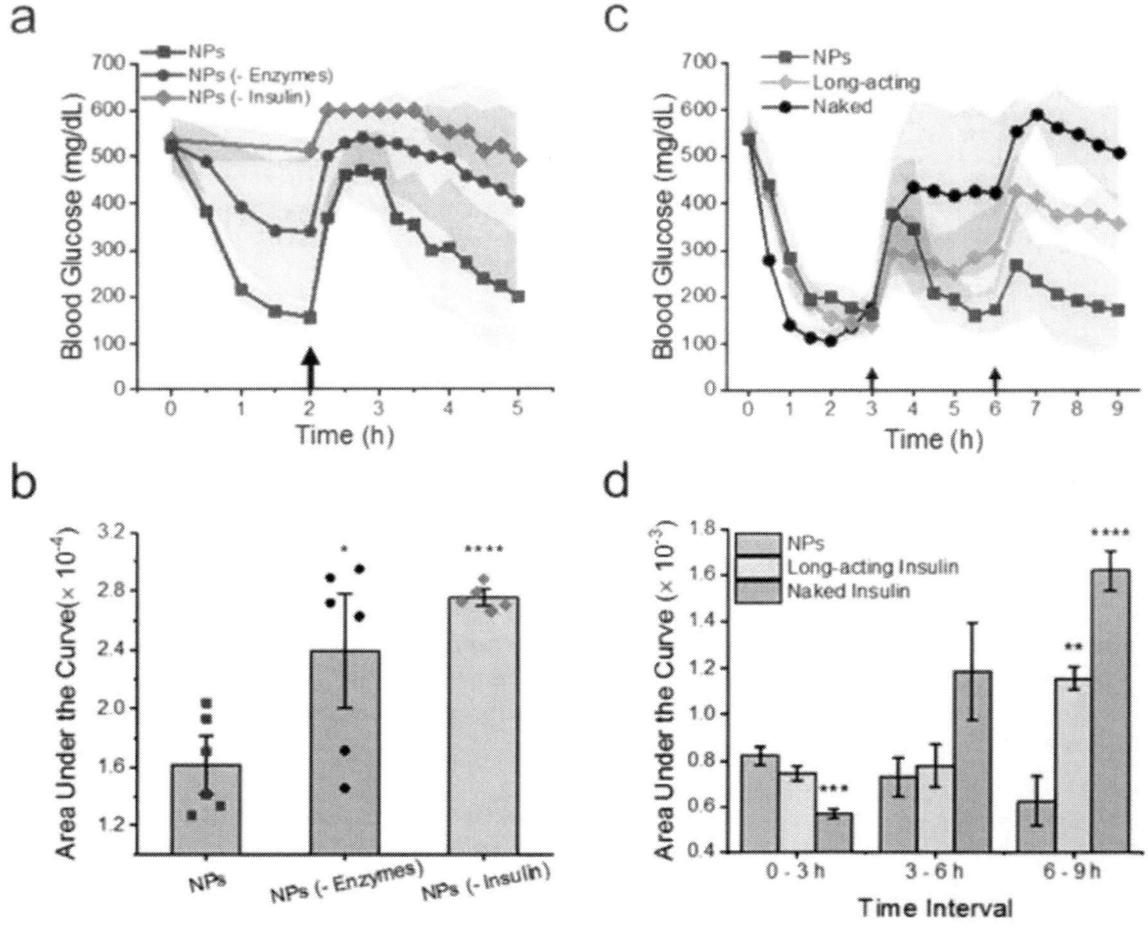

Fig. 1. Glucose-responsiveness provides enhanced glycemic control. a) Response of diabetic mice to a glucose injection (3 g/kg, indicated by arrow) following administration of 9 IU/kg NPs, 9 IU/kg NPs lacking enzymes, or an equivalent dose of NPs lacking insulin. b) Area under the curve for (a). c) Blood glucose levels of diabetic mice following administration of 5 IU/kg NPs, long-acting insulin, or naked insulin with glucose injections at 3 h and 6 h (1.5 g/kg). d) Area under the curve for (c) by 3 h time interval. Shaded regions represent standard deviation. Error bars represent standard error of the mean (N = 4+). Statistical significance is indicated by *$P < 0.05$, **$P < 0.01$, ***$P < 0.001$, ****$P \leq 0.0001$ in comparison to the first NP bar (blue).

Methods:
GOx acts as a glucose sensor by catalyzing the reaction of glucose to gluconic acid, consuming oxygen and producing hydrogen peroxide as a byproduct. Therefore, polymers that respond to changes in pH, oxidation, or hypoxia may be used in conjunction with this enzyme for self-regulated insulin delivery. Here, acetalated-dextran (Ac-dex) nanoparticles (NPs) were formulated containing insulin, GOx, and catalase. Ac-dex rapidly degrades as a result of the reduced pH from the GOx reaction when glucose levels are high, thus releasing insulin on demand. Catalase provides added biocompatibility by disproportionating the hydrogen peroxide byproduct.

Results: To distinguish between glucose-responsive and non-responsive insulin release, diabetic mice were administered nanoparticles containing enzymes and insulin (NPs), nanoparticles without enzymes (NPs (– Enzymes)), or nanoparticles without insulin (NPs (– Insulin)). Following a glucose injection, only the NP group was able to regain glycemic control within 3 h (Figure 1a). Therefore, the glucose sensor and the insulin are necessary to produce consistent glycemic control with significantly reduced areas under the curve (AUCs, Figure 1b). Diabetic mice were found to have the best glycemic control when administered

NPs, compared to an equivalent dose of a long-acting acylated analog of insulin (commercially known as insulin detemir) and naked insulin (Figure 1c). Specifically, for the time interval between 6 – 9 h, the long-acting insulin and naked insulin groups both have significantly higher AUCs than the NP group (Figure 1d). Therefore, for the same dose of insulin, glucose-responsive NPs provide better extended glycemic control compared to long-acting or naked insulin treatments.

Conclusions: The enzymatic sensor glucose oxidase can provide a trigger for applications in self-regulated insulin delivery. The resulting glucose-responsive nanoparticles provide better glycemic control than a system without such sensing capabilities, which may improve diabetes therapies in the future.

576890 In Vivo Biosensing Platform for Detection of Complex Human Milk Oligosaccharides

Human milk oligosaccharides (HMOs) are prebiotic compounds found in breast milk that favor the growth of beneficial bacteria in the infant gut. While there are many efforts underway to synthesize these beneficial complex carbohydrates, the current standards for their detection are the relatively low-throughput chromatographic methods or mass spectrometry. We present a high-throughput, fluorescence-based detection system for HMOs in vivo in *E. coli* (Enam and Mansell, *Cell Chemical Biology,* 2018). Such high-throughput systems facilitate directed evolution and enables a streamlined synthesis process for prebiotics with a significant reduction in analysis time of new producer strains. We also present combinatorial approaches towards construction of elements for detection of more complex HMOs. Additionally, the reporter enables the use of HMOs as orthogonal inducers of cellular circuits in synthetic biological applications.

564854 Protein Detection with Peptoid-Functionalized Carbon Nanotube Optical Sensors

A limitation to real-time imaging of metabolites and proteins has been the selective detection of biomolecules that have no naturally-occurring or stable molecular recognition counterparts.[1] We present developments in the design of synthetic near-infrared fluorescent nanosensors based on the fluorescence modulation of single-walled carbon nanotubes (SWNT) with select sequences of surface-adsorbed *N*-substituted glycine peptoid polymers.[2] We assess the stability of the peptoid-SWNT nanosensor candidates under variable ionic strengths, solution composition, protease exposure, and cell-compatible conditions, and find that the stability of peptoid-SWNTs depends on the composition and length of the peptoid polymer. We identify a peptoid-SWNT assembly that can detect lectin protein wheat germ agglutinin (WGA) with a sensitivity comparable to the concentration of serum proteins. To demonstrate the retention of nanosensor-bound protein activity, we show that WGA bound to the nanosensor produces an additional fluorescent signal modulation upon exposure to the lectin's target sugars, suggesting the lectin protein remains active and selectively binds its target sugars through ternary molecular recognition interactions relayed to the nanosensor. Additionally, we demonstrate that peptoid-SWNT assemblies present a generic platform for protein detection, by demonstrating extension of peptoid-SWNTs for detection of additional protein analytes. Our results inform design considerations for developing synthetic molecular recognition elements by assembling peptoid polymers on SWNTs, and also demonstrate these assemblies can serve as optical nanosensors for lectin proteins and their target sugars. Together, these data suggest peptoid sequences can be assembled with SWNTs to serve as versatile optical probes to detect proteins and their molecular substrates.

1) Landry, M. P., ... Chio, L., *et al.* Single-molecule detection of protein efflux from microorganisms using fluorescent single-walled carbon nanotube sensor arrays. *Nature Nanotechnology* (2017). doi:10.1038/nnano.2016.284

2) Chio, L., Del Bonis-O'Donnell, J.T., Kline, M.A., Kim, J.H., McFarlane, I.R., Zuckermann, N., Landry, M.P. Electrostatic-Assemblies of Single-Walled Carbon Nanotubes and Sequence-Tunable Peptoid Polymers Detect a Lectin Protein and its Target Sugars. *Nano Letters,* (2019) DOI: 10.1021/acs.nanolett.8b04955

564279 Label-Free QCM Immunobiosensor for Real-Time Detection of GFP Antigen Using IOS-1 Peptoid:

The medical diagnostic industry is in need of cheaper, time-saving, and more robust assays for pathogen detection. Our research approach is focused on the development of short-chain designer peptoids that can be tuned to target detection of specific biological molecules. The need for detection of specific pathogens has necessitated our design of a designer peptoid (IOS-1), which preferentially isolates a target antigen (i.e., pathogen or pathogen metabolite) from a dynamic aqueous stream. When incorporated into diagnostic tools, these designer peptoids, which are customizable to the extracellular matrix (envelope) protein of target pathogens, may proffer a cost-effective and rapid means of pathogen identification and quantification. In this research, the precise Quartz Crystal Microbalance (QCM) is used to track adsorption events in real-time.

In this work, we have developed a technique for the in situ detection of green fluorescent protein (GFP) antigen in flowing PBS buffer solution using an immunobiosensor comprising a gold substrate whose surface is modified ex-situ with IOS-1 peptoid. This peptoid is an oligomer of N-substituted glycines that was designed using the submonomer solid-phase synthesis approach.

The QCM provides real-time frequency shift data for the mounted immunobiosensor during the flow of analyte solution, baseline solution, and reflux solution. By using the Sauerbrey model, mass loading of the adsorbed analyte was calculated, thereby enabling in situ quantification of adsorbed GFP antigen. This immunobiosensor is robust enough for dynamic operation and is reusable in vitro by switching the flow of antigen with 0.1% sodium dodecyl sulfate solution to recondition the immunobiosensor surface before resumed capture of antigen. Equilibrium mass loadings of antigen before and after SDS reflux point toward reproducible surface chemistry for this immunobiosensor. Experimental data obtained using this setup are contrasted with data from a similarly constructed immunosensor consisting of the gold substrate which is surface modified with aminoethane thiol, disuccinimidyl suberate and GFP antibody in a stepwise immobilization scheme.

The average mass loading of peptoid immobilized on the (0.78 cm^2) sensor surface was about 400 nanograms, and this peptoid monolayer can capture between 1 nanogram of the analyte and 200 nanograms of the analyte, depending on the initial concentration of antigen flowed across the immunobiosensor. Characterization of the surface was achieved using FTIR, fluorescence microscopy and XPS. Baseline QCM readings were taken in PBS buffer.

574737 High-Throughput Assaying of Individual Host-Pathogen Dynamics in Influenza A Virus Infection Using Drop-Based Microfluidics

Influenza A virus (IAV) is notorious for causing seasonal epidemics and sporadic worldwide pandemics that have resulted in over 100 million deaths during the past century. Understanding how IAV interacts with its host cells is typically performed using tissue culture systems or with animal models. These systems provide bulk measurements of virus production and/or cellular responses, such as interferon activation and production. However, individual cells exist as a heterogeneous population, resulting in cell-to-cell differences in virus production and antiviral responses. Conventional methods for measuring the antiviral response in individual cells during IAV infection are labor intensive, incur high reagent costs, and require extensive serial dilutions to isolate singularly infected cells. Thus, there is a need for high-throughput analysis of single cells to elucidate individual host-cell dynamics. Here, we have developed a drop-based microfluidic method to characterize IAV infection and quantify the cellular response to viral infection at the single cell level. This method utilizes drop-based microfluidics to encapsulate infected cells within picoliter sized drops followed by single cell, drop-based qRT-PCR analysis. We successfully formed $>10^6$ drops that contained individual human alveolar epithelial (A549) cells infected with the pandemic 2009 H1N1 IAV. Our results show that IAV replicates within the drops and that drops remain stable overnight with high cell viability. We quantified viral RNA and cellular mRNA from these infected drops at kilohertz rates using our novel, multiplexed drop-based qRT-PCR assay. This approach allowed us to measure interferon ß production in response to viral RNA output from $>10^5$ individually infected cells in a single experiment. Our method is the first to measure individual host-pathogen dynamics in single cells following infection with IAV using a high-throughput, multiplexed drop-based assay.

564163 Minimally Invasive Extraction of Plant DNA *Via* a Polymeric Microneedle Patch for on-Site Detection of Plant Pathogens

Molecular diagnoses play an essential role for identification of plant pathogens and global crop protection. However, extraction of DNA from plant cells is challenging due to the presence of surrounding polysaccharide cell walls. As a result, isolation of high-quality plant DNA is mainly confined to well-equipped laboratories, and sample preparation becomes one of the major hurdles to perform molecular diagnostic assays of plant pathogens in the field. To expedite sample preparation and pathogen detection on site, a simple and rapid DNA extraction method has been developed using a polymer microneedle (MN) patch to isolate genomic or pathogenic DNA from plant leaves. This minimally invasive DNA extraction method isolated amplification assay ready DNA from several plant species within a minute. Furthermore, MN extraction of pathogenic DNA was tested and compared to conventional cetyltrimethylammonium bromide (CTAB) extraction by isolating *Phytophthora infestans* DNA from infected tomato leaves for late blight disease detection. For laboratory-inoculated samples, the MN extraction matched the performance of the CTAB method for samples after 3 days of inoculation. For field-collected samples, MN patches successfully isolated *P infestans* DNA from all leaf samples. These MN patches could be used to diagnose other plant pathogens from various plant species, and thus have great potential to become a general sample preparation technique for on-site molecular diagnosis of plant diseases.

576177 Nanostructured Polymer Interfaces for Improved Lectin-Based Capture and Detection of Bacteria

There is critical need for developing materials and interfaces that can be used to capture and isolate bacteria for rapid, culture-free sensing at high sensitivity. Lectin interfaces are considered to be promising for affinity-based microorganism capture and isolation from complex samples, including blood, urine, and wastewater. However, there are limitations associated with the practical use of many lectin-based interfaces due to the fact that the equilibrium dissociation constants (K_D) of lectin-carbohydrate interactions are higher than antibody-antigen binding constants by 2-3 orders of magnitude, resulting in poor cell capture efficiency.

In this work, we have designed surfaces that combine reactive polymer coatings that offer control of biomolecule surface density with nanoscale surface structures that allow for precise control of physical surface topology in order to improve cell capture. First we characterize the critical interface parameters that impact lectin coupling onto flat, reactive azlactone-based block copolymer films that consist of poly(glycidyl methacrylate)-*block*-poly(vinyldimethyl azlactone) ($PGMA_{56}$-*b*-$PVDMA_{175}$). Second, we incorporate well-controlled physical nanostructures into the films using nanospike arrays (black silicon) fabricated from reactive ion etching, which provide underlying physical structures for the polymer coating. The microbe detection sensitivity and bactericidal impact of the both flat and nanopillar polymer surfaces are then characterized.

Our data indicate that the physicochemically-optimized surfaces enhance capture and detection of bacteria from solutions by two orders of magnitude compared to control surfaces, which were lectins immobilized to flat surfaces using standard carbodiimide coupling chemistry (EDC/NHS). We found that flat PVDMA surfaces alone improve the limit of detection (LOD) by one order of due to the fact that the PVDMA polymer offered higher levels of lectin surface density. Addition of the nanopillar arrays to the PVDMA-coated surfaces improved the sensitivity by an additional order of magnitude, down to 10^2 CFU/mL. This is likely due the three-dimensional surface structures that provided enhanced contact with the targeted bacteria. This level of sensitivity gives our interfaces relevance to a variety of bacterial detection applications which include detection in water/wastewater systems (required LOD =10^3-10^5 CFU/mL), urinary tract infections (required LOD = 10^3 CFU/mL), and blood infections (required LOD = 1-100 CFU/mL).

579458 The Valence of Skills: The Rise of Hybrid Jobs

Abstract TBD
New technologies are creating new jobs, but they are also changing the skills required for existing jobs. Overall, the half-life of skills is declining rapidly. But a more profound change is the way in which skills are bonding with one another in unexpected and sometimes transformative ways. These hybrid jobs challenge traditional, siloed structures of education. What's more, in this environment, technical skills, while critical, are not sufficient—even in STEM roles. To keep from becoming obsolete, today's students will need to acquire a set of New Foundational Skills, combining technical expertise with the human skills to work with others and the business skills needed to get things done.

580071 Ethical Dilemmas - an Introduction

This introduction will suggest different types of ethical considerations and will outline ethical dilemmas typically found in engineering education, government and business. In addition, results will be shared from the 2017 Ethics Survey of AIChE Members.

580126 Data: Yours? Mine? Ours? and When Is Enough Enough?

As engineers, we handle lots of data, interpret data and recommend actions based on data all with an ethical underpinning of safety for the public. We must consider regulations, business objectives, privacy and security concerns and more which can combine with human and organizational paradigms and objectives to create pressure or conflicts that result in ethical dilemmas. Our first case study will put participants in real life situations relevant to both the research environment and industry.

580127 Ethical Dilemmas in Medical Devices: How much clinical data should be required to approve a medical device?

This presentation will explore the ethical dilemmas that can arise in the design, development, approval and marketing of medical devices.

580074 Workshop: Dealing with Ethical Dilemmas

Working groups will be formed in this breakout session to allow attendees to work on realistic case studies that face engineers in education, research and business.

580076 Post-Workshop: Discussion

In this session, the groups will come together and share their efforts to handle the ethical dilemmas presented in the case studies and challenges they face to get to this decision.

588861 The Chemical Engineering Profession and Its Infinite Possibilities

Tayo Femi-Fowode is a passionate Chemical Engineer with over 10 years of corporate and leadership experience. He is currently a Business Integration Leader for GE Power's Monitoring and Diagnostic Center in Atlanta, GA. In his role, Tayo drives digital integration of 5000 monitored turbines, generators and other equipment to minimize downtime, optimize plant performance and save operating cost for more than 950 power plants in 75 countries and serving 350 million people worldwide. Before joining GE Power in 2011, he was Project Manager in the asset management division of Rolls-Royce Energy.

A second-generation Nigerian-American, Tayo holds an MBA from Columbia Business School with concentration in Strategy and Finance. He also holds a Master of Science and a Bachelor of Science *(magna cum laude)* degrees in Chemical Engineering from Columbia University and Prairie View A&M University respectively.

588862 Direct contact membrane distillation (DCMD) membrane distillation techniques for desalination

Direct contact membrane distillation (DCMD) is one of the simplest and most widely used membrane distillation techniques for desalination. It is a thermally driven separation process and is sought to achieve high water vapor flux while being thermally efficient. Typically, a hydrophobic porous membrane is used and water vapor from the salt solution is allowed to pass through the pores and condense on the other side of the membrane. Investigation on other structures of membranes is carried out. Membranes include those in the form of an integral composite, in which the structures of PVDF membranes are changed, and a stacked composite, in which different membranes are stacked on top of each other (e.g. a hydrophobic ePTFE membrane over a hydrophilic PVDF membrane, and a hydrophobic PVDF membrane over another hydrophobic PVDF membrane). These membranes were found to yield a higher water vapor flux than the conventional single hydrophobic membrane. Membranes are also characterized using a variety of characterization techniques: Contact angle, liquid entry pressure (LEP), bubble-point pressure, scanning electron microscopy (SEM), and atomic force microscopy (AFM).

588863 My Career Path

My career thus far has been shaped by many factors learned early in life. It started in my hometown of Yazoo City, Mississippi. It was there that I had the job that I consider to be my first engineering experience which was working with my dad at Gainwell Tire Services. While there, I discovered that I had a fondness for hands-on work and gravitated toward working closely with people as part of a team. This was instrumental in not only my early career development but shaped my life as a whole. From there, while in high school, I worked at our local construction company, Parker Builders. During this time, I developed a strong partiality for repairing process equipment. Also, a sense of work ethic was formed as there were several days where I worked on a hot tin roof for hours. After graduation, I headed to college at Louisiana Tech University. While there, I was a part of the football team while pursuing my Chemical Engineering degree. This conceivably impossible task was made possible largely due to the work ethic learned early in life. It also helped further developed my team-membership skills.

Later, I transferred to Southern Arkansas University. It was here that I not only compounded upon my team-membership skills but developed into a team leader as a captain of the football team. While there I took on several internships which included stints in the chemical, aerospace, energy, and mechanical fields. One of those internships was Lanxess Solutions in El Dorado, Arkansas. The skills learned early in life helped me very much in many of the projects given during this time and were good enough to be hired full time as a Shift Leader upon graduating college. In this role, I developed many technical skills and gained an understanding of the operations side of the plant. After some time there, I went to New Orleans, Louisiana to work at Galata Chemicals as a Process Engineer. While at Galata, I was able to further develop technical skills as well as learn the safety aspect that is imperative in a plant setting. After spending a year and a half there, I was offered a Process Engineer II position at Oxea Chemicals, my current role. Thus far I have learned even more about the chemical field and continue to grow professionally with the organizational training offered.

588932 Sustainable Energy Drivers: A Perspective from a Career in the Nuclear Sector

Dr. Harold T. Conner, Jr. has a 52-year career in the nuclear sector, having determined at an early age that he wanted a career in chemical engineering. Along the way, he blazed an impressive path by achieving a lot of firsts: the first African American student in the University of Tennessee-Martin Engineering School, the first cooperative engineering student at the University, and the first African American to receive bachelor's and master's degrees in chemical engineering from the University of Tennessee-Knoxville. He worked for more than 30 years in Oak Ridge, working in almost every facility at the Oak Ridge K-25 Site, a uranium enrichment facility that would be closed down in the mid-1980s and then undergo cleanup. Conner became site manager of the K-25 Site and was then named vice president of Environmental Management and Enrichment Facilities. He would leave Oak Ridge and worked at various DOE complexes across the nation until returning to Oak Ridge to help with the cleanup of the K-25 Site, which is now called the East Tennessee Technology Park (ETTP). Working for UCOR, an AECOM led partnership with Jacobs, his expertise has significantly helped the company work toward achieving Vision 2020, DOE's plan to complete ETTP cleanup by 2020. The site is being transformed into a private sector industrial park, national park, and conservation area. Conner is now semi-retired but serves as a consultant for UCOR.

588933 A Chemical Engineer in the Energy Journey

Dr. Quinta Nwanosike Warren, PE, PMP is a chemical engineer and an energy professional. She is an entrepreneur and has worked in industry, academia, and government. She holds a PhD in Chemical & BioMolecular Engineering from the Georgia Institute of Technology, and a Bachelors in Chemical Engineering from the Pennsylvania State University. She is a Project Management Professional, and a registered Professional Engineer.

Dr. Warren previously worked for ConocoPhillips for on CO2 capture research, and as a reservoir engineer for heavy oil and tight gas assets. She has worked with the US Department of Energy, learning about and contributing to domestic and international policy around carbon management and power generation. She also worked for the Millennium Challenge Corporation (MCC), a US government agency that works on international development. At MCC, she provided energy expertise and contributed to the development of energy projects.

Dr. Warren owns and runs a consulting company called Energy Research Consulting (www.EngrRC.com) which provides energy and water solutions and strategies. The company's projects include creating the STEM-focused Dakar American University of Science and Technology (www.DAUST.org) in Senegal, and designing a water treatment system for a mushroom farming initiative in Nigeria. Dr. Warren recently published a textbook titled "A Practical Guide to Oil & Gas Resource Characterization for Geologists and Reservoir Engineers."

581829 Negotiation and Evaluating a Job Offer for Nerds

Networking is all about building win-win partnerships that last a lifetime, where both parties provide value to each other. Networking is the most important tool you have in your professional advancement toolbox because it is the only means by which you can access hidden jobs or perhaps even create your dream job from scratch. But most people don't realize that networking begins, ends, and indeed requires an attitude of generosity, where I am looking to see how I can help you and add value, not get something from you. In this presentation, we will discuss what exactly networking is, and how to actually "do" networking effectively with this spirit of generosity that enables everyone to achieve their goals. We will also address specific networking tips you can employ at the AIChE Annual Meeting!

580230 Workshop--Part 1 "I Followed the Recipe, Where Is My Cake?": Important Questions to Ask in Considering Transition to Management

You are an individual contributor, and you aspire to become a leader. But have you thought about why you want to change to management and how one makes that transition? Most people deliberate over it for a long time and work towards this goal. The roles you take in the years before you want to make the jump will demonstrate that you are ready for management. This session will focus on steps to take and opportunities to target to prepare for the transition into management.

580231 Workshop--Part 2: I Know the Why. Now What?

If you know the why, the next thing on the list is to figure out what needs to be done to show that you are ready to be a manager. What should you be prepared for? Unfortunately, they don't teach us how to be a manager in school, and degrees don't necessarily land us a management job. In this workshop, we will list down some activities and opportunities that we can take to be able to demonstrate that you are prepared for management roles.

580232 General Discussion

The workshop leaders will summarize and open the session for general discussion.

581830 Engineers Can Do Anything: Explore Your Passions and Catalyze Your Potential to Create Your Unicorn Career

We all face challenges at some points in our careers, whether we have been laid off, need to leave our current position due to either positive or negative stimuli, are looking for a promotion, or are eager to work in a completely different field or profession. But no matter the challenge, engineers have the competitive advantage to manage, survive and thrive because of your innate problem-solving skills. This presentation will address how to manage a career challenge and tackle it using your strengths as an engineer. We will explore options for handling especially challenging situations, like zombie bosses (who steal your work or malign your reputation) or a desire to make a mid-career change, and chart a specific, strategic course to achieve your goals.

588881 Using microcontact printing to mimic Cardiomyocyte regeneration

Cardiac diseases are the most prominent cause of death worldwide. Cardiomyocyte regeneration is imperative, due to the human body lacking in the ability to sufficiently regenerate cardiomyocytes (CMs). One-way cardiomyocytes can be regenerated is through differentiating human induced pluripotent stem cells(hiPSCs) into cardiomyocytes. In order to attach hiPSCs derived-cardiomyocytes in a manner that would mimic native conditions, various techniques can be used, such as microcontact printing. The microcontact printing technique allows for a specified geometry or pattern to be transferred onto the substrate that will be used for cardiomyocyte seeding. The pattern utilized in this study comes from the rectangular-pillared-array stamp. The preliminary results show that a general cell line of mouse embryonic fibroblasts (MEFs) can be constrained to this pattern.

588862 Direct contact membrane distillation (DCMD) membrane distillation techniques for desalination

Direct contact membrane distillation (DCMD) is one of the simplest and most widely used membrane distillation techniques for desalination. It is a thermally driven separation process and is sought to achieve high water vapor flux while being thermally efficient. Typically, a hydrophobic porous membrane is used and water vapor from the salt solution is allowed to pass through the pores and condense on the other side of the membrane. Investigation on other structures of membranes is carried out. Membranes include those in the form of an integral composite, in which the structures of PVDF membranes are changed, and a stacked composite, in which different membranes are stacked on top of each other (e.g. a hydrophobic ePTFE membrane over a hydrophilic PVDF membrane, and a hydrophobic PVDF membrane over another hydrophobic PVDF membrane). These membranes were found to yield a higher water vapor flux than the conventional single hydrophobic membrane. Membranes are also characterized using a variety of characterization techniques: Contact angle, liquid entry pressure (LEP), bubble-point pressure, scanning electron microscopy (SEM), and atomic force microscopy (AFM).

588882 Effects of a functionalized CNC on the rheological properties and printing behavior (Direct Ink Writing (DIW) of an epoxy resin

In this work we characterize the effects of a functionalized CNC on the rheological properties and printing behavior (Direct Ink Writing (DIW)) of an epoxy resin. CNC functionalization is achieved by grafting an activated medium chain fatty acid to substitute the hydroxyl functional groups. The level of CNC modification is evaluated using Fourier Transform Infrared (FTIR), thermo-gravimetric analysis (TGA), X-ray diffraction (XRD), and solvent dispersibility studies. Functionalized CNCs are dispersed into the epoxy resin (EPON 828) at loadings up to 20 wt%. Initially, rheological studies are used in this study to assess the printability of the inks. The rheological percolation threshold is the critical concentration at which there is a transition from liquid-like to solid-like behavior. For this system, this value was found to be at ~10wt%. We found that for the final printed structure to exhibit structural integrity, it is necessary to print at loadings slightly above this value. Future work will include microscopy (SEM and polarized light) of both printed and cast samples to observe the dispersion and alignment of the CNCs within the polymer matrix. Finally, the mechanical properties of all samples will be measured to show the effect of the reinforcement in the matrix.

588883 madison.garner1@my.hamptonu.edu

Conventional lithium batteries using an organic polymer electrolyte, exhibit several safety issues including leakage, poor chemical stability and flammability. To address these safety concerns the idea of a solid electrolyte has emerged. In the lab we specifically focused on the solid garnet-type electrolytes. The solid electrolyte of $Li_{6.28}Al_{0.24}La_3Zr_2O_{12}$ and $Li_{7.28}Al_{0.24}La_3Zr_2O_{12}$ (LALZO) powder were successfully synthesized by a co-precipitation method. LALZO pellets were then sintered at 1100 °C at different sintering times of 4,8, and 16 hours. With the lithium content of 7.28 per formula, impurity of La_2O_3 was obtained at the sintering time of 4 hours. When the sintering time increased to 8 and 16 hours, in addition to the La_2O_3, $La_2Zr_2O_7$ phase was obtained. When the lithium content decreased to 6.28 per formula, impurities of La_2O_3 and $La_2Zr_2O_7$ were obtained. The lattice parameters are all similar for different lithium content and different sintering times and are within the range of 12.93-12.97Å.

588874 Epitranscriptomics of RNA oxidation reveals mechanisms of lung toxicity and disease

Chemical modifications of proteins and nucleic acids play critical roles in the regulation of gene expression. New evidence suggests that environmental stresses may be involved in the misregulation of the functional activities of RNA modifications and pathways during lung distress and disease. Air pollution is the major source of environmental toxicity to the pulmonary cells. Components of air pollution can generate free radicals and perturb redox homeostasis in cells, leading to formation of the prominent RNA oxidative modification 8-oxo-7,8-dihydroguanine (8-oxoG). This mark is predominantly induced in respiratory cells, and it leads to genotoxicity as it preferentially mispairs with adenosine. Given the prevalence of RNA oxidation under environmental challenges, it is critical to identify novel molecular mechanisms involving RNA oxidation toxicity to cellular physiology and disease. To address this, we developed a high-throughput sequencing method to profile 8-oxoG in mRNAs of human bronchial epithelial BEAS-2B cells. Cells were exposed for 1.5 hours to realistic air pollution mixtures derived from the reaction of ozone (100 ppb) with acrolein (100 ppb), methacrolein (97ppb) and a-pinene (44ppb), common organic volatile compounds in the atmosphere. These compounds formed a multi-component gas-phase mixture and submicron secondary organic aerosols (PM 1.0) with concentrations ~ 40 $\mu g/m^3$. Using this approach, we found 42 transcripts that are consistently oxidized by air pollution. These transcripts belong to key oxidative stress, signaling, and metabolic pathways. Importantly, one of the mRNA transcripts that is more susceptible to oxidation is FDFT1, a protein that catalyzes a regulatory step in cholesterol synthesis and free-radical scavenging. Moreover, we confirmed a ~2-fold (p-value < 0.05) decrease in expression of FDFT1 protein, and a ~1.7-fold (p-value < 0.001) decrease in cellular cholesterol levels. To further validate the involvement of misregulation of cholesterol on cellular toxicity of air pollution, we knocked down FDFT1 using anti-sense siRNA in BEAS-2B cells. We found that FDFT1 knockdown promotes distinct morphological phenotypes typically observed in studies of lung stress and inflammation. Overall, our findings indicate that air pollution influences the formation of 8-oxoG marks in transcripts of epithelial lung cells leading to alterations on cellular function and cell morphology.

588871 Cell-to-cell variations define age-dependent cell migration through fractional redistributions among motility states

As humans age, cells within organs and tissues undergo profound biophysical and biomolecular changes, which significantly influence the rate of progressive functional decline. Recently we have demonstrated that aging information is encoded within biophysical properties of cells, and as such can be used as biomarkers of aging in health and disease. The coordinated movement of cells within organs and tissues are important in the contexts of both homeostasis and disease; such as in the surveillance of immune cells and during cancer metastasis. Although it is now recognized that there are ensemble decreases in the motile capacity of cells with increasing age, it is unclear how this motile regulation becomes impaired. To address this question we procured a panel of primary dermal fibroblasts from donors aged 2-92 years old, and tracked the movement of cells seeded on collagen-coated substrates for ten hours, taking images every 3 minutes. Quantification of bulk motility properties using the anisotropic persistent random walk model, revealed that there was indeed a decrease in the overall capacity of cells to move with increasing age (including displacement, directionality and velocity). Looking further into these findings, we plotted the trajectories of individual cells across all ages, and found that there was a high decree os cellular heterogeneity among cells, with some cells from elderly donors exhibiting highly-motile features, and vice versa. Prompted by the extent of cell-to-cell variations, we hypothesized that these age-dependent decreases in cell motility was not due to global decreases in the movement of all cells, but a re-distribution of the proportions of highly-motile and non-motile cells with increasing age. Taking advantage of the single-cell nature of the data, we compiled the trajectories of cells across all ages and computed eight motility parameters that described the spatiotemporal movement; e.g. velocity, displacement, anisotropy and diffusivity. Then from this data matrix we used unsupervised hierarchical clustering and identified eight clusters which defined 'motility states' exhibiting distinct spatial patterns, as indicated by the trajectories of individual cells within each cluster, and t-stochastic neighbor embedding analysis. Next, we asked whether certain clusters had increased propensities for cells of a particular age. Quantifying the abundance of cells classified per cluster for all ages, we found that there was an increased propensity of various cluster for young and elderly donors, however, for post-adolescent/adult donors there was a similar likelihood of either cluster. This finding was interesting as it points to increased heterogeneity during the 'middle-age' period. To further verify this result, we quantified the 'activity' of cells per age, as determined by the 1- dimensional displacements per cell based on the frequency and magnitude of displacement bursts. Utilizing a similar clustering approach as before, we identified five clusters that defined the activity of each cell across all ages. interestingly, while the temporal patterns of the bursts were different among groups, plotting the trajectories of individual cells we were unable to identify distinct spatial motility patterns. However, plotting the abundance of each 'activity' cluster per age, showed a less-defined but similar tendency in cellular heterogeneity. Combining the information regarding the spatial-temporal and activity per cell, we computed the abundance of cells within the 40 motility states (8 spatio-temporal clusters x 5 activity clusters) to define motility landscapes, and asked how this differed as a function of age group. Interestingly, we found that there was a clear difference among age groups regarding the propensity for certain states, with the significant gain and loss of states across ages. Quantifying the Shannon entropy of each landscape, we confirmed the results of a maximization of heterogeneity for donors classified as post-adolescent/adults.

588870 Microfluidics for the Engineering Transport Classroom: Multipath Flow

Microfluidic devices can be utilized to recapitulate a myriad of engineering principles. While polydimethylsiloxane (PDMS) chips are the most common, hybrid polyethylene terephthalate laminate (PETL) microfluidic devices chips have shown promise in being cheaper, more affordable, and faster alternatives to the current "gold-standard". In an effort to apply newer, more hands-on methods to assist in teaching engineering principles in the classroom, PETL devices were successfully implemented in modeling multipath flow in microfluidic systems. Future developments may allow for more complex means of integrating other engineering concepts for higher accuracy and a wider scope of application.

578803 Solvent Absorbent Amphiphilic Microgels for Pickering Emulsions

Pickering emulsions are critical to systems where encapsulation of an oil or water phase in an immiscible medium is required. These coated emulsions can harbor oils in water or even water in oils. While hard particles are traditionally used, microgel particles have also been used as emulsion stabilizers. These polymer colloids differ from rigid particles in many ways. When compared to microgels, colloidal particles are lyophobic, meaning they need surface modification to be preferably "wet" by a solvent. Microgels can be made to be lyophilic, having an attraction to the solvent that swells them. Microgels are cross-linked polymer networks that absorb and desorb solvents where the type of solvent used to swell a microgel can be tailored during particle synthesis.

Using a Polydimethylsiloxane microfluidic device, via soft lithography, the flow streams of two separate polymer solutions were brought into contact to form Janus droplets. The individual droplets were then exposed to UV light to induce cross-linking via photo-polymerization to form monodispersed microgel particles with sectioned hydrophilic and hydrophobic sides. 45 wt% Polyethylene Glycol diacrylate (PEGDA), 5 wt% ethoxilated trimethylopropane triacrylate (ETPTA), 1 wt% 2-hydroxy-2-methyl-propiophenone (HMP), and 49 wt% water were mixed together to form the hydrophilic portion while 99 wt% Polypropylene Glycol diacrylate (PPGDA)) and 1 wt% 2-hydroxy-2-methyl-propiophenone are added together to form the hydrophobic side. In crosslinking these two systems together, we form single particles that have one side capable of absorbing water and another with the ability to absorb and desorb organic solvents. These particles were used to make stable water in oil and oil in water Pickering emulsions.

PEG functionalized CdSe/ZnS core-shell type quantum dots in water and CdSeS/ZnS alloyed quantum dots in toluene were used to observe the absorption abilities of the hydrophilic and hydrophobic microgel sides respectively with a confocal microscope. While oriented at the water/oil interfaces, the respective microgel sides swell and capture the dispersed nanocrystals. These amphiphilic microgels pose as an interesting tool in stabilizing emulsion droplets for cosmetic, pharmacy, and petroleum industries.

588868 Determining the Effects of Adipose-derived Exosomes on Stem Cell Proliferation and Adipogenesis

Determining the Effects of Adipose-derived Exosomes on Stem Cell Proliferation and Adipogenesis

The field of tissue regeneration has made many strides to improve therapeutic methods and restore missing functions to patients with the introduction of stem cell therapies. Transplantation of fat tissue has provided beneficial metabolic effects to patients by supplying sufficient vascularization and innervation[1]. Despite the benefits that adipose tissue therapeutics provide, there persists limitations such as the restriction to autologous therapies due to immunogenic responses. Exosomes may serve as a novel cell-free method to provide the same therapeutic properties while circumventing potential immunogenic responses. This is beneficial to patients who are deficient in fat content and would not have tissue for an autologous therapy, such as lean and pediatric patients. The purpose of this project is to determine the therapeutic potential of adipose-derived exosomes on tissue regeneration and their influence on human adipose-derived stem cell (hASC) adipogenesis.

Fig 1. Schematic of growth factor effects on adipogenic gene expression[4]

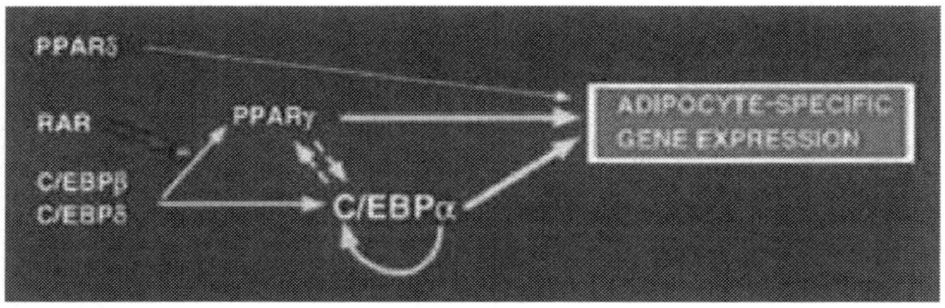

Exosomes are small vesicles (40-100nm) that are secreted by a variety of cell types into the extracellular matrix. Within an apoptotic cell, the intracellular matrix fuses with the plasma membrane to produce multivesicular bodies (MVB). Through fusion with the plasma membrane, exosomes retain the properties that are associated with the native cell. These vesicles have surface ligands that help to facilitate cell-to-cell interactions without the need for direct cell contact. By inducing the expression of target growth factors, we can interact with the various surface proteins that are involved in cell proliferation and differentiation[2]. Exosomes fuse with target cells to exchange membrane proteins and cytosol to signal intended functions[3]. The surface proteins of cells provide a more accessible route for communication than internal targets. Transcription factors that have been found to induce adipogenesis include members of the CAAT/enhancer binding protein (C/EBP) and peroxisome proliferatoractivated receptor (PPAR) families (Figure 1). Through a combination of interactions, each transcription factors can directly or indirectly activate the expression of adipogenesis. C/EBPß and C/EBPd have synergistic effects to influence PPARγ and C/EBPa, each of which have a direct influence on gene expression. PPARd also has a direct influence on adipocyte gene expression, with slightly less significance[4].

By directing the surface interactions, it is possible to have a cell perform specific tasks, in response to the given environment. **I hypothesize that adipose-derived exosomes can serve as a therapeutic agent by promoting proliferation and differentiation.**

Currently, exosomes from differentiating human skeletal muscle cells have been tested to observe the regenerative properties within skeletal muscle tissue. Studies in other tissues have also shown that mesenchymal stem cells (MSC) derived exosomes have functions such as repairing tissue damage, suppressing inflammatory damage, and modulating the immune system[5]. Working with adipose-derived tissue, I want to determine the therapeutic potential that exosomes possess. Future work will consist of determining the mechanism behind each function. This therapy has the potential to assist with tissue regeneration in patients that are dealing with adipose tissue defects. Exosomes show promise for transferring regenerative therapeutics between patients without having to encounter immunogenic rejection, a major setback when dealing with stem cells. This research will open new avenues for dealing with patients who lack the tissue mass for autologous regeneration therapeutics such as lean or pediatric patients. Restoring functions to damaged organs can one day be achieved by mediating the repair of the desired tissue through exosome mediated cell-to-cell signaling.

588867 Predicting Cancer Progression using a Population Balance Model and Supporting Evidence from Zebrafish Melanoma Studies

Cancers are complicated diseases that in 2018, killed 600,000 people in the US. Reliable models for the progression of the disease save lives by guiding treatment and identifying risk factors. Of special interest is the perplexing phenomena of tumor dormancy and recurrence, where after treatment or surgery, patients remain disease-free for years or decades only to experience a sudden relapse of the same disease. We seek to explain this and other phenomena with a simple population balance model that describes how growth effects (e.g. mitosis), reduction effects (e.g. immunity or treatment) and metastases (inception of new tumors) determine disease progression.

In the model, these effects are represented as tumor-size-dependent rate parameters in a stochastic process. Treating cancer growth as an advection-diffusion transport process in tumor size space allows us to solve for the size distribution of tumors in a population of patients. Simulations using the model show that tumor size can change diffusively for long times after which the tumor begins to grow or shrink, indicating the model recapitulates the phenomena of dormancy and recurrence without any manipulation.

We used the zebrafish melanoma model to quickly acquire tumor growth data under various conditions including growth with and without the natural immunity and growth in different genders. The size-dependent parameter approach described experimental data well. We also saw statistically different immunity parameters between male and female fish, indicating how the model can be used to identify and elucidate population differences in disease progression. We further show through simulations backed by experimental data how changes in parameters affect the possibility of relapse using both parameters from animal and human cancers.

The model demonstrates a mechanism for tumor dormancy and recurrence that does not rely on any specific biological explanation such as the so called 'angiogenic switch.' Rather it shows that a series of small, incremental changes such as those due to a size-dependence can generate dormancy and recurrence. Recognizing that these phenomena could have no single biological cause mandates a radical shift in research methodology for this problem.

588866 Research Explore Removing Zinc Oxide from the coastal marine environments prone to excessive sunscreen runoff.

With recent highlights of the state of coral reefs and help from social media, the public has finally become aware of the adverse consequences of the widely used chemical, Oxybenzone, in conventional sunscreen, as well as its direct effect on coral bleaching. The issue has become so prevalent in coastal areas, that a prohibition of Oxybenzone will take action in Florida and Hawaii in 2021 (Calderone, par. 1-2). To support the movement, consumers have been taking it on their own behalf to choose "coral reef safe" sunscreens. The two prominent ingredients in the eco-friendly sunscreen movement are Zinc Oxide and Titanium Dioxide. Recent studies, however, point to evidence that these two chemicals may not be as Eco-friendly as producers are suggesting.

Zinc Oxide in sunscreen has a tendency to dissolve into Zinc ions in water, and with decreasing the particle size, there is an increase in the likelihood of releasing free ions. The effects of Zinc ions in water has been shown to promote the growth of algae. Additionally, ZnO showed increased effects in solutions that were aged, suggesting that the true effects of the chemical won't be apparent until later on.

On the other hand, initially, it may seem intuitive to pursue Titanium Dioxide sunscreen. Unfortunately, Titanium Dioxide is a 2B carcinogen, meaning that while it is not a confirmed fact yet, it may cause cancer to humans (Smijs, par 21).

Additionally, new generation organic sunscreens, while they have been on the market for the past few years in the EU, cannot be used yet as the compounds they contain have not yet been approved in the U.S. by the FDA.

588865 Spray Drying of Griseofulvin Nanosuspensions and Solutions for Preparation of Nanocomposites and Amorphous Solid Dispersions: Comparative Assessment of Drug Release

BCS class II drugs are compounds that are poorly water-soluble and exhibit high permeability. In order to increase the dissolution rate and bioavailability of these compounds, two specific processes can be used: amorphous solid dispersions (ASD) and nanocomposites. ASD's disrupt the drug crystal lattice which causes an increase in the saturation solubility, whereas nanocomposites exhibit very large surface areas; both of which cause a faster dissolution. The goal of this project is to compare the release of the active pharmaceutical ingredient Griseofulvin (GF) from nanocomposites and that from ASD's, both having identical formulations.

GF nanocomposites were prepared by wet media milling (nanomilling) an aqueous GF suspension followed by spray drying. Two different polymers, hydroxypropyl cellulose (HPC) and Soluplus, with an anionic surfactant, sodium dodecyl sulfate (SDS), were used for the stabilization of drug nanoparticles. ASD's were formulated by spray drying the respective drug-polymer in acetone-DI water. The suspensions were characterized via laser diffraction, while the spray-dried powders were characterized via laser diffraction, X-ray diffraction (XRD), and dissolution testing. XRD diffractograms confirmed that for both polymers, spray drying of the aqueous nanosuspension led to nanocomposites, whereas that of the drug solution led to ASD. At supersaturating conditions, drug dissolution from ASD was faster than that from the nanocomposites.

588864 Properties of Modified Kraft Lignin for Use in Thermoplastic Composites

With the ever-increasing use and popularity of thermoplastics, effective waste management continues to be an area of interest. In order to preserve or improve the quality of the recycled material, thermoplastic composites have been researched. The properties of modified Kraft lignin by two methods (i.e. acetylation and maleation) are investigated and the results are compared with virgin Kraft lignin. Kraft lignin is considered a commercial byproduct of the Kraft pulping of wood chips. In this study, the thermal and morphological properties of the esterified Kraft lignin are analyzed. The lignin is modified by esterification using acetic and maleic anhydride to alter solubility behavior in nonpolar solvents for application in thermoplastics. The morphology of the composite materials is examined using scanning electron microscopy (SEM). Differential scanning calorimetry (DSC) and thermogravimetric analysis (TGA) is used to study the glass transition temperature and thermal stability of the modified lignin samples. The results are compared with virgin Kraft lignin.

580685 From Artificial Cofactors to Synthetic Arrays: Teaching Old Enzymes New Tricks

Modern biotechnology comprises remarkable capabilities to engineer biological systems and impart them with novel functions, from living organisms with reprogramed genomes to individual proteins with optimized structures. Another resource and relatively recent addition to the biotechnology armory are the extremophiles, which span the limits of life on earth and represent a vast but still largely untapped source of new bioproducts and biomaterials. For proteins in particular, opportunities to expand their range of function and/or create new activities have never been greater, and include protein engineering by directed evolution or rational design, harnessing proteins from extremophiles, and equipping enzymes with unnatural chemical moieties that enable them to carry out abiotic transformations. Accelerated rates of catalysis, enhanced stability against denaturants, higher binding affinities of ligands, and even programmed self-assembly and designed mechanical functions are currently possible and rapidly advancing with improved techniques and understanding of protein structure-function relationships. Application areas that we have focused on recently include expanding biocatalysis to include abiological reactions and reaction conditions at the outer limits of life, and developing advanced biomaterials with programmable functions. Progress on each of these fronts will be discussed.

588877 *2019 Eminent Engineer Recipient:* Ana Davis, Syngenta: "What I know now that I didn't know then – A Chemical Engineer's Journey"

Ana will reflect on her journey as chemical engineer and touch on the "breakthrough" moments that have led to positive changes in her life and career as a chemical engineer. She will touch on topics such as Diversity, Leadership, Empowerment and Courage. She will share personal stories that have helped her grow as a leader and as an individual.

588878 *2019 Eminent Engineer Recipient:* Zenaida Otero Gephardt, Rowan University - "Talent and Hard work are essential for success"

Talent and hard work are essential for success, but without mentors and meaningful support, success can elude even the most talented and industrious among us. This is especially true for those who are underrepresented and have historically lacked access to opportunity. The presentation will focus on the importance of mentors throughout the presenter's career and on how mentoring is an essential investment in the chemical engineering profession and in the global community. As mentors, we see reflected the best of ourselves and continuously improve in our professional and personal lives regardless of career stage. The presenter will also discuss her personal experience with the type of support that is necessary to overcome poverty and disadvantage. Mentoring is key to the broadening of participation in the profession that isessential to identify opportunities in future challenges, and to implement solutions that improve quality of life worldwide.

588880 *2019 William W. Grimes Award:* Dr. Christine Grant, NCSU- "Mentors, Mentors Everywhere: Weaving Informal and Formal Mentoring into a Robust Chemical Sciences Mentoring Quilt"

In an era of convergence, a clear understanding of the discipline-specific cultures is at the root of interdisciplinary mentoring. Informal and formal mentoring manifests differently for women and underrepresented minorities in the chemical sciences where more work is required to achieve parity, equity, and inclusion for diverse groups. This is particularly true for Women of Color (WOC) who occupy a unique place at the intersection of race and gender in STEM fields. Intentional, informed, mentoring includes informal interactions at formal conference events, where cross-cultural mentoring can result in robust, long-lasting mentoring teams.

581346 Response of Magnetic Nanoparticles to Time Varying Magnetic Fields: Ferrohydrodynamics, Nanoscale Thermal Therapy, and Magnetic Particle Imaging

Colloidal suspensions of magnetic nanoparticles are fascinating from an intellectual and practical perspective because of their behavior in externally applied static and dynamic magnetic fields. This includes particle translation in magnetic field gradients, internal dipole or whole-particle rotation when the local magnetic field changes direction, and dissipation of magnetic field energy in the form of heat, among others. Suspensions of magnetic nanoparticles in Newtonian fluids, commonly known as ferrofluids, have been of interest since the 1960's due to their practical applications and unique fluid mechanics. More recently, biomedical applications of magnetic nanoparticles have been of great interest. Magnetic nanoparticles have been clinically translated for use as magnetic resonance imaging contrast agents, in sentinel lymph node detection, and in nanoscale thermal therapy for cancer. Furthermore, the nonlinear magnetization response of superparamagnetic iron oxide nanoparticle tracers is critical for magnetic particle imaging, a new molecular imaging modality that allows for non-invasive, tomographic, unambiguous, and quantitative imaging of tracer distribution. This talk will focus on the response of magnetic nanoparticles to applied time varying magnetic fields. First, the phenomenological description of ferrofluid flows in rotating magnetic fields will be presented, along with theoretical and experimental work suggesting the existence of antisymmetric and couple stresses, features of structured fluid continua. Second, experimental evidence will be presented in support of nanoscale thermal phenomena in the vicinity of magnetic nanoparticles, resulting in lysosomal permeabilization and death in cancer cells without a macroscopic temperature rise. Third, magnetic particle imaging guided nanoscale thermal therapy will be presented, along with recent results demonstrating unprecedented spatial resolution imaging using strongly interacting magnetic particle imaging tracers.

580905 Introductory Remarks for the Industry 4.0 and Digital Transformation Session

The distinguished industrial speakers will be introduced by the session chair.

579773 Leading the Industry 4.0 Journey in the Manufacturing Sector with Cross-Industry Collaboration and AI-Powered Innovation

While manufacturing represents 16% of global GDP, manufacturing industries account for 64% of global R&D spend. Yet innovation is long overdue since the manufacturing sector has experienced a decade of productivity stagnation and demand fragmentation. Organizations that have taken Industry 4.0 innovation to scale beyond the pilot phase have experienced unprecedented increases in efficiency with minimal displacement of workers. However, most companies appear to be stuck in "pilot purgatory". Widespread adoption of artificial intelligence (AI) and allied Industry 4.0 technologies at scale, through the combined efforts of companies and governments, can lead to a sizeable increase in global wealth creation, benefiting people throughout society.

However, there is potential for worker displacement if changes are not handled properly. Strengthened collaboration among stakeholders is needed to understand unexpected consequences and manage the transition.

Leaders in digital manufacturing and globalization 4.0 exhibit all the essential characteristics of the Industry 4.0 revolution. Moreover, they confirm the hypothesis that they have the potential to generate new economic value, driven by improvements in the full spectrum of production value drivers: resource productivity and efficiency, agility and responsiveness, speed to market and customization to meet customer needs. Transforming existing production systems, innovating value chains and building new business models with disruptive potential for established businesses can create value.

579424 Real-Time Analytics for IIoT

Digital transformation, connected enterprise, and IIOT are various names that refer to an increasing interest by the manufacturing industry to make better use of the data they generate. A key driver behind this heightened interest is the expectation that the business will benefit from mining actionable information out of that data.

The prevailing discourse today equates successful digital transformation with big data analytics (often delivered via a cloud-centric infrastructure). This talk will argue that, at least in industrial applications, there are a significant subset of problems for which a cloud-centered big data approach is not necessary.

This talk will present a family of embedded analytics capabilities that are designed to extract actionable information out of streaming operation data. A brief description of each capability along with the results from real-world applications of the technology will be presented.

579314 Process Optimization in the Age of Industry 4.0

Industry 4.0 represents a new trend of automation and data exchange in manufacturing technologies and fosters the concept of a "smart factory". Within this framework cyber-physical systems monitor physical processes, create a virtual copy of the physical world and make decentralized decisions. This talk provides an overview of modern optimization optimization tools and solution strategies within this distributed decision-making process. In contrast to current engineering tools which are based on simulation models and expand to optimization, we describe the development of more advanced decision-making tools that start with an optimization model and flesh out the details of the process. In particular, we highlight the development of equation oriented optimization platforms with fast solvers, advanced modeling environments and efficient interfaces between them. Moreover, we extend these features to NLP sensitivity, nonsmooth models and data-driven models. Finally, we explore the application of these tools for distributed optimization, optimization under uncertainty and the integration of design, operations and control.

580904 Value-Driven Digitalization Initiatives at Dow

Largely, the process industry has matured beyond the initial frenzy around automated reports, dashboards, and the embedding of statistics in software products, and has firmly moved toward building out infrastructure and decision-support systems to improve safety and maximize value.

Industry megatrends around M&A for narrowed domain focus, commoditization, low capital intensity, geo-political uncertainty, and emphasis on sustainability and customer experience require nimble and responsive supply chains, systems analysis for understanding trade-offs, and maximization of value from asset footprints. Further, with the emergence of AI platforms, the time has never been more ripe for practitioners to help an allegedly sluggish and slow-to-adopt industry to embed digital tools as a core part of their strategy.

A lot of emphasis has rightly been on external visibility of product portfolios to customers, as well as on internal decisions around what, when, and where to produce product---consequently involving IT, customer service/marketing, and supply chain functions. Although there continues to be significant opportunity in this space, AI technology and analytics has become increasingly relevant to R&D groups in the development of new products and innovative processes; and to maintenance and operations organizations in predicting and responding to process events.

Further, most digitalization projects fall in one of three areas of increasing implementation complexity: (1) technology that enables improved decision-making in a local environment that can be then cloned in different environments with small tweaks; (2) technology that is specific to a particular domain, business, class of products, or process; or (3) technology that spans across multiple assets and allows consolidated decision-making at scale. These three categories differ in the effort needed to standardize, scale, and get buy-in from stakeholders. These pose slightly different challenges, and the way organizations structure support organizations, empower employees, and manage programs is critical to the successful deployment of these systems and in effecting culture change.

Here, we provide specific examples where Manufacturing 4.0-related projects have enabled product, process, maintenance, and supply chain innovation at Dow, including recent success stories of deploying process intensification at scale. We also cover what Dow has done to intensify its organizations through three key capabilities---the Digital Marketplace, the Digital Operations Center, and the Digital Fulfillment Center, where domain experts and engineers sit alongside software architects, UX designers, and database experts to deliver solutions effectively. Further, we discuss how communities of practice around analytics have helped us align analytics professionals across different functions to develop a coherent strategy for the company around, for example, deployment best practices, building data pipelines, and digital twins. External partnerships with universities, consultants, and vendors do play a key role in technology transfer and adoption, and balancing the development of in-house talent with the right external support is important as well.

Through these stories, we hope to illustrate how an organization can plan a steady and firm climb out of the trough of disillusionment with digital technologies into the plateau of productivity.

580906 Q&A Panel Discussion

The distinguished speakers will hold a Q&A session with the audience, moderated by the session chairs.

580292 Background and Discussion about Retaining Women Engineers

Dr. Urmila Diwekar is currently President of the Stochastic Research Technologies LLC, and the Vishwamitra Research Institute (VRI, www.vri-custom.org), a non-profit research organization that she founded in 2004 to pursue multidisciplinary research in the areas of Optimization under Uncertainty and Computer-aided Design applied to Energy, Environment, and Sustainability.

From 2002-2004, she was a Professor in the Departments of Chemical Engineering, Bio-Engineering, and Industrial Engineering, and in the Institute for Environmental Science and Policy, at the University of Illinois at Chicago (UIC). From 1991-2002 she was on the faculty of the Carnegie Mellon University, with early promotions to both the Associate and the Full Professor level.

In chemical engineering, she has worked extensively in the areas of simulation, design, optimization, control, stochastic modeling, and synthesis of chemical processes. She has made major contributions to research on batch distillation, and this work is well recognized. She is the author of more than 165 peer-reviewed research papers, 7 books, and 15 chapters, and has given over 350 presentations and seminars, and has chaired numerous sessions in national and international meetings. In separations, she is well-known for her pioneering work in batch distillation design, simulation, optimization, and optimal control.

She has published more than 31 peer-reviewed journal papers, 65 presentations (15 plenary/keynote presentations), and conducted 11 educational workshops on this topic. She has also published the first book on this topic. This is an authoritative, first of its kind book which is widely used all over the world. This book received excellent reviews and the second edition of this book was published in early 2012. She is the author of a comprehensive software package for batch distillation simulation, optimal design, and control called MultiBatchDS which is used in pharma and specialty chemical companies in USA, Europe, and Asia.

In this session, she will give her personal opinion about how to maintain women in the industry.

580293 Background and Discussion about Retaining Chemical Engineering Women in Industry

Gayle J. Gibson led the DuPont Engineering Research and Technology organization of internal engineering consultants and the company's Field Engineering Program for early-career engineers.

Gayle joined DuPont in 1983 and held a variety of roles through which she implemented best practices and enabled global business revitalization on behalf of multiple, high-performing business units. Her contributions included engineering design, process engineering, research and development, product and process development, manufacturing supervision, market development, business strategy, growth and supply chain improvement.

From 2008-2013, Gayle was Director – Corporate Operations for the Chair of the Board of Directors and Chief Executive Officer of DuPont. In that role she facilitated processes and engagement among key DuPont operating and leadership teams. This included working across businesses, functions and regions to drive alignment and clarity on priorities, advancing core values and resolving issues in line with the company's mission and strategic objectives. She also supported the Chief Executive Officer as co-lead of the National Academy of Engineering project "Changing the Conversation – from Research to Action" to improve the public understanding of engineering.

A certified Six Sigma Champion, Gayle holds a Bachelor of Science in chemical engineering from Texas A&M University and a Master of Business Administration from the European University. She is a member of AIChE, Tau Beta Pi and the Society of Women Engineers

In this session, she will be giving her personal thoughts on how to retain women engineers in the industry.

580291 Background and Discussion about Retaining Women in Industry

Ana Davis heads Health, Safety, and Environment (HSE) in North America for Syngenta, a global agriculture company. She manages all of Syngenta's corporate HSE programs, which include active ingredient manufacturing, product formulation, seeds operations, and commercial sales. Her team develops tools to improve safety culture across the company.

Ana began her career as a process control engineer with Exxon, later working as a facilities engineer for Advanced Micro Devices. Prior to joining Syngenta in 2009, she held a variety of production and HSE management roles with Kao Specialties Americas. Before that, she was a plant manager at Dow and held several positions with Union Carbide Corp., becoming a plant production manager after the Dow Chemical merger.

A native of Spain, Ana was recognized in 2005 with the Hispanic Engineer National Achievement Award for Professional Achievement in Industry. She serves on the advisory boards of AIChE's Center for Chemical Process Safety (CCPS) and North Carolina State Univ.'s Dept. of Chemical and Biomolecular Engineering. Ana received her BS degree in chemical engineering from North Carolina State Univ., where she and her husband have established an endowment for chemical engineering undergraduate students.

In this session Ana will give her personal point of view about maintaining women engineers in the industry.

587477 Background and Discussion about Retaining Women Engineers in Industry

Bill serves as Director – Engineering Technology for the Chemours Company. He is accountable for leading the company's efforts to transform and renew the leveraged engineering competencies in support of Manufacturing Operations, large Capital Growth Projects, Sustainability and Process Safety.

Bill has over 30 years of experience with DuPont and Chemours in R&D, Engineering and Manufacturing technology management, as a well having assignments in corporate risk management and trade secret protection.

Bill graduated with a B.S. degree in Chemical Engineering from Columbia University and received his M.S.C.E.P. and Ph.D. degrees in Chemical Engineering from the Massachusetts Institute of Technology.

581347 Chemical Engineering Innovations for a Renewable Economy

Chemical engineering evolved in the twentieth century with rise in the use of fossil resources and myriad of products derived from then. An array of these products required great innovations and made huge impact on the human race, and such examples include urea, transportation fuels, polymers, water purification chemicals, etc. Going forward as we make transition from fossil resources to renewable resources such as solar energy, wind and sustainably available biomass, we again have great opportunities for innovation that would not only make great impact but redefine chemical engineering. My talk will focus on my own experiences with challenges and opportunities presented to us as chemical engineers. I will share an overall perspective on the use of biomass as sustainable source of carbon, harvesting solar energy in conjunction with growing food without any land constraint, and creating solution processed solar cells that utilize chemical engineering principles along with knowledge from other engineering disciplines.

581350 Engineering Materials for Clinical Application

Dr. Schmidt's research has resulted in numerous patents that are licensed to companies, including AxoGen, Alafair Biosciences, Smith and Nephew, and Siluria Technologies. Comments for this panel will focus on peripheral nerve applications and one key success that is the foundation for the Avance Nerve Graft from AxoGen. In addition, along the way to uncovering strategies for nerve regeneration, Dr. Schmidt and her team also discovered a biomaterials processing approach that led to a separate application of protecting tendon's after surgery, resulting in the VersaWrap Tendon Protector from Alafair Biosciences, a start-up company in Austin, Texas.

Nerve Repair: Dr. Schmidt's technology for processing decellularized nerve tissue was licensed to AxoGen in Alachua, Florida (http://www.axogeninc.com) for the AVANCE™ nerve graft. Many tens of thousands (~100,000) of AxoGen's AVANCE™ grafts have been implanted in patients at hundreds of hospitals globally.

Post-surgical Wound Management: Using a novel in situ crystallization processing technique, Dr. Schmidt's team created mechanically robust hyaluronic acid-based hydrogel films that can be easily handled, sutured, and delivered laparoscopically. This research is the basis of Alafair Biosciences, a start-up company in Austin, TX (http://www.alafairbiosciences.com), which is run by Dr. Schmidt's PhD graduate and a neurosurgeon collaborator. Alafair's first product, VersaWrap™ Tendon Protector, has been implanted in more than 500 patients since December 2017.

579453 Control of Interfacial Stability in Electrochemical Systems

The shape and composition of electrochemical interfaces often undergo instabilities, which are notoriously difficult to understand and control in engineering applications. The first part of this talk will describe three examples of interfacial patterns – viscous fingers, deionization shocks, and metal dendrites – whose stability can be controlled by electrokinetic phenomena in charged porous media, as evidenced by both theory and experiments. Potential applications include electrically enhanced oil recovery, water desalination and purification by shock electrodialysis, and energy storage with rechargeable metal batteries. The second part of the talk will describe how driven electrochemical reactions can alter the thermodynamic stability of solid or liquid interfaces, with applications to Li-ion batteries, electrodeposition, corrosion, and biological pattern formation.

580356 My Management Journey, from Technology Developer to Innovation Leader

An innovation leader embodies domain expertise, perseverance, risk-tolerance, creativity, adaptability and trustworthiness, with a strong scientific and engineering foundation. Innovation leaders believe the right science applied in the right way can impact people worldwide, generating breakthroughs that make life better. At 3M, my teams leverage 51 technology platforms to solve customers' problems. I have experienced first-hand the joy of discovery, the power of curiosity and the importance of determination to deliver new technologies and products to diverse markets and improve lives. I will describe experiences and moments of pride for my team's achievements.

581071 Impacts on Management from an Evolving World and Profession

Chemical engineering and the world are evolving; how does that affect management? In some essential ways, not at all: A decision is still a decision, a smile is still a smile as time goes by. Communicating the organization's mission clearly is still essential. So is rewarding people who perform well, not the ones who don't.

In other ways, everything has changed: Staff are expected to be reachable instantly, travel for meetings is slashed in favor of virtual "face-to-face" and audio meetings, and detailed management data on budgets, supply chains, and production are available instantaneously for analytics and decision-making - at least in principle. None of that is chemical-engineering-specific.

The profession-driven changes for chemical engineers are more subtle. The pace of mergers and company re-organizations is due in part to diverse opportunities in material, bioscience, and information technologies. Similarly, employees themselves have more diverse expectations. Perhaps more than in any other engineering discipline, education of chemical engineers is to learn a core of expertise, not a specific job. In recent years, students also learn more about the diverse ways they'll apply their education, and they often look forward to being entrepreneurial. That presents opportunities and challenges for their managers.

581072 Advancing Technology Though Measurement Science

The mission of the National Institute of Standards and Technology (NIST) is to promote U.S. innovation and industrial competitiveness by advancing measurement science, standards, and technology in ways that enhance economic security and improve our quality of life. We do this by working in close partnership with US industry to build the technical infrastructure needed for advancing technology. In this presentation, I will highlight successful examples pf this partnership from the nSoft consortium and the Materials Genome Initiative.

581073 Engineering and Science at the National Laboratories

The National Laboratories represent complex organizations that conduct forefront research, and that house some of the most advanced technological instruments available to the scientific community. In this presentation, I will present a brief overview of Argonne National Laboratory and of Fermilab, and how industry and academia interact with these institutions to pursue large multidisciplinary projects involving large teams of scientists and engineers. Importantly, the success of these projects depends on robust management structures that have evolved over the years to enable and facilitate scientific research. I will describe some of these structures, along with some of their advantages and limitations.

580360 General Question and Answer Period

The Chairs will moderate a general Q&A between the audience and Dr. Thomas and the other speakers.

580948 Colloidal Self-Assembly of Mechanically-Reinforced Organohydrogels

There has been a surge of interest in composite hydrogel materials that can be spun into fiber form, because of advantages associated with increased moisture holding, mechanical enhancement, and biocompatibility. Nevertheless, a major challenge is that it is difficult to generate fibers with functional microstructure without expensive post-processing steps. We introduce a colloidal gel system comprised of nanoemulsion droplets of poly(dimethylsiloxane) (PDMS) suspended in a continuous phase comprised of a double network hydrogel precursor that contains sodium alginate and poly(ethylene glycol diacrylate) (PEGDA). The nanoemulsions undergo self-assembly at elevated temperatures to form hydrophobic networks with different length scales, which can be crosslinked by UV light due to its hydrogel phase. We find that the length scales of the colloidal nanoemulsions are tuned independently using the applied shear rate and temperature in a cylindrical spinneret, generating solid organohydrogel fibers with different microstructures. Furthermore, we show that it is possible to introduce a double network hydrogel precursor that is compatible with the self-assembling nanoemulsions, thus generating fibers with enhanced Young's modulus, fracture energy, and rupture stress. This study utilizes state-of-the-art understanding of colloidal gel rheology and hydrogel chemistry to generate new types of mechanically reinforced materials that can also be used to control drug loading and release.

580949 Translocation of Cell-Penetrating Peptides into Fungal Cells

Cell-penetrating peptides (CPPs) are small peptides that can cross cellular membranes and carry cargo into cells. Although the interactions of CPPs with mammalian cells have been widely studied, much less is known about their interactions with fungal cells. We have studied the translocation of a number of peptides into the human fungal pathogen *Candida albicans* using a combination of simulations and experiments. We have shown that the mechanisms of translocation vary by peptide and include both endocytic mechanisms and direct translocation. To study how net charge and hydrophobicity of CPPs affect translocation into fungal cells, we designed variants of the peptides pVEC and SynB with altered levels of charge and hydrophobicity and evaluated translocation into *C. albicans*. Charge played a greater role in translocation efficiency of the peptides than hydrophobicity, with a higher level of charge leading to higher level of translocation into *C. albicans* and a higher level of cytosolic localization. Hydrophobicity had little effect on translocation efficiency, but a low level of hydrophobicity did lead to increased vacuolar localization and an energy-dependent translocation mechanism. Our results suggest that CPPs can be designed for desired levels of cargo delivery into fungal cells and for desired translocation mechanisms, which will provide useful insight as new CPPs are designed to specifically target fungal cells.

580950 Organic Solvent Reverse Osmosis Separations of Hydrocarbons Using "Polymer-Derived" Membranes

The rapid increase in global industrialization necessitates technology shifts in energy production, manufacturing, and carbon management techniques. Approximately 10-15% of global energy use can be attributed to separation processes, with the vast majority of separations being "thermal" in nature (e.g., distillation). Membranes are potentially disruptive technologies, but issues of perceived risk/reliability, scalability/cost, and performance must be addressed for membranes to move beyond water and gas separations and towards world-scale sectors such as hydrocarbon processing. The polymer processing platform for the creation of membrane devices is eminently scalable and has the potential to match the enormous volumes associated with chemical and petrochemical separations. "Polymer-derived" organic solvent reverse osmosis (OSRO) membranes with the capabilities of separating hydrocarbon compounds of similar size and shape have recently emerged and will be the focus of this talk. Exemplar "challenging" separations in the hydrocarbon processing space will be discussed, and potential membrane-enabled solutions will be provided. Finally, a comparative performance analysis between OSRO membranes based on pure polymers, organic-inorganic hybrids, and "polymer-derived" carbon membranes will be given.

580952 Global Optimization of Stiff Dynamical Systems

In this talk, a new deterministic global optimization method is presented for nonlinear programming formulations constrained by stiff systems of ordinary differential equation (ODE) initial value problems (IVP). The examples arise from dynamic optimization problems exhibiting both fast and slow transient phenomena commonly encountered in model-based systems engineering applications. The proposed approach utilizes unconditionally-stable implicit integration methods to reformulate the ODE-constrained problem into a nonconvex nonlinear program with implicit functions embedded. This problem is then solved to global optimality in finite time using a spatial branch-and-bound framework utilizing convex/concave relaxations of implicit functions constructed by a method which fully exploits problem sparsity. The algorithms were implemented in the Julia programming within the EAGO.jl package and demonstrated on numerical examples with varying complexity relevant in process systems engineering. The developed methods enable the guaranteed global solution of dynamic optimization problems with stiff ODE-IVPs embedded.

580953 Impact of Surface Loading on Catalytic Activity of Regular and Low Micropore SBA-15 in the Knoevenagel Condensation

The immobilization of highly active catalysts on heterogeneous catalytic materials has the potential to enable the efficient and sustainable production of chemicals and fuels. Yet, catalyst immobilization on mesoporous supports such as SBA-15 typically results in low catalytic activity. Herein, we demonstrate that a key limitation for heterogeneous catalytic materials is the presence of micropores. Through reducing the micropore volume, we are able to increase the catalytic activity by over a factor of two for the Knoevenagel condensation and other related catalytic reactions. This observation has the potential to overcome the key limitation of reduced catalytic activity for immobilized catalysts, encouraging the broad adoption of these promising catalytic materials.

580951 Machine Learning Predictions of Electronic Couplings for Charge Transport Calculations of P3HT

Improving the inexpensive generation of electricity with organic semiconductors requires an understanding of how charges move through these materials. Here we aim to predict charge mobility in films of poly-(3-hexylthiophene) with varying degrees of molecular order. We lower the computational cost of these predictions by training machines to determine electronic couplings between polymer chromophores, which is a bottleneck in this prediction pipeline. We evaluate five machine learning techniques and find random forests accurately predict couplings 390 times faster than performing quantum chemical calculations from scratch. Using these couplings as inputs to kinetic Monte Carlo simulations of charge transport we find zero-field hole mobilities within 5% of prior work. We conclude with discussions of the training data that are needed to inform these accurate predictions and preliminary applications of this approach to more complex organic semiconductors.

580300 Part 1 – Applying Creativity Tools

In **Part One** of the workshop (approximately 75 minutes), you will learn how to jumpstart your creativity with a simple exercise that yields a high number of ideas for further exploration.

580301 Part 2 – Practicing Master Minding

In **Part Two** of the workshop (45 minutes), we will test drive a sample advisory board by brainstorming solutions to the challenges introduced in Part One.

580302 Part 3 – Setting & Achieving Goals

Finally, we will conclude with **Part Three** (30 minutes) of the workshop in which you'll learn how to design achievable goals with accountability from your leadership advisory board.

With a mix of learning and hands-on practice, you will learn how to apply five leadership mindsets to help you achieve higher levels of collaboration within your teams while building a toolkit for fast-tracking your career success. Moreover, these techniques will help you balance work, leisure, health, and relationships so that you will feel more joy and purpose as a multi-dimensional chemical engineering leader. If you are interested in accelerating your career as a leader in your organization, this workshop is for you!

581105 Systems Biology of Metabolism : Role in Production of Advanced Biofuels, Obesity and Cancer

Metabolism is the core of functioning of any cell as it ensures provision of Gibbs free energy as well as precursors for synthesis of cellular constituents like proteins, lipids and DNA. Metabolism involves a large number of biochemical conversion processes. Thus, even Baker's yeast, that serves as the most simple model for studying human cells, contains more than 900 enzymes that catalyze more than 1,500 biochemical reactions. In human cells these numbers are much larger with more than 3,000 enzymes and more than 5,000 biochemical reactions. Even though the large number of reactions are organized into metabolic pathways, there is a high degree of connectivity between the reactions, and hence it is quite complicated to study these reactions individually. It is therefore necessary to take a systemic approach for analysis of metabolism, often referred to as systems biology. We are working on generating so-called genome-scale metabolic models (GEMs) that are comprehensive description of cellular metabolism. We have over the last years reconstructed GEMs for a number of industrially important fungi, including the Baker's yeast *Saccharomyces cerevisiae*, and used these models for analysis of large data sets and for identification of novel targets where we can engineer the metabolism, often referred to as metabolic engineering. Hereby we have developed advanced cell factories for the production of fuels and chemicals. Recently we have also build a Human Metabolic Atlas, a novel web-based database and modelling tool that can be used by medical and pharmaceutical researchers to analyse clinical data with the objectives of identifying biomarkers associated with disease development and improving health care. The central technology in the Human Metabolic Atlas is GEMs, which are tissue-specific. These models allow for context-dependent analysis of clinical data, providing much more information than traditional statistical correlation analysis, and hence advance the identification of biomarkers from high-throughput experimental data that can be used for early diagnosis of metabolic related diseases. In this presentations our technologies behind reconstruction, simulation and analysis of GEMs will be presented and results from studies in metabolic engineering and systems medicine will be presented. In connection with the latter it will also be discussed how we can advance towards modeling of the gut microbiome, which has recently demonstrated to be an active metabolic organ in the human body.

571808 Remaining True to Yourself

The sudden transition from being a Postdoctoral Researcher, working under a principal investigator, to becoming an Assistant Professor leading an own group, can be a daunting balancing act, somewhat akin to starting a business, but with the unique human dimension associated to working in education.

There is a wonderful sense of purpose, with vastly increased independence and the opportunity to mentor, but the journey, navigating the complex and ever-evolving space of academia, is also full of challenges: from realizing a personal vision and leadership to the daily operational elements; from preparing and teaching new courses to setting up a lab; from raising funds for research, to tutoring and advising students. And, as the ultimate challenge: Remaining true to yourself and maintaining a healthy balance in life.

I will share some thoughts, based on my international and US-based experiences, from Assistant Professor to Head of Department, via professorships in the Netherlands, the USA and the UK.

566789 Reflections from Ten Years at the Lectern: Insights on Teaching, Scholarship, and Service from an Educational-Focused Perspective

If asked to identify what are the three most important elements of a professor's life, a common reply might be, "teaching, scholarship, and service." As faculty members progress through their careers, it is important to identify how they want to make an impact in these areas. It is understandable that different types of universities place different emphasis on these three elements: an R1 research university might put more emphasis on research, whereas an primarily undergraduate institution (PUI) might put more emphasis on teaching. But what about service? Service can mean many things: service to your department, service to your institution, and service to your profession.

In this talk, I will provide insights into advancing teaching, scholarship, and service from the perspective of working at an undergraduate-focused institution. More specifically, I will focus primarily on teaching and service. The teaching part of my talk will focus on adopting modern pedagogical methods and leveraging resources to enhance student learning in the classroom. Examples of teaching initiatives that I have worked on include peer instruction, digital textbooks, and incorporating computation in the classroom. The service part of my talk will provide insight into how to identify opportunities to make a difference both on an institutional and professional society level. Examples of service initiatives that I have worked on include serving on multiple search committees, chairing my university's Faculty-Student Senate, leading institutional accreditation efforts, serving as a volunteer ABET evaluator, and working in leadership roles at AIChE and ASEE.

While I will provide some background on the opportunities that I have pursued in the areas of teaching, scholarship, and service, ultimately the goal of this talk is to provide advice to young faculty members in achieving success and ideally a valued sense of fulfillment in these three important areas of faculty life.

579451 From Chemical Engineering Fundamentals to the Commercialization of Vapor Deposited Polymers

Fundamental considerations enabled the invention of mechanistically-driven processes for chemically vapor deposited (CVD) polymers. Utilizing selective chemistry allows CVD polymer growth to proceed rapidly on room temperature surfaces with full retention of the polymer's organic functional groups. The initiated CVD (iCVD) and oxidative CVD (oCVD) methods mimic chain-growth and step-growth polymerization, respectively. Adsorption of the vapor phase reactants is often rate-limiting and follows the Brunauer Emmett Teller (BET) adsorption isotherm. With this knowledge, iCVD and oCVD homopolymer and copolymer thin films has been demonstrated for nearly 100 monomers.

CVD can be the only viable single-step film forming method for polymers which are incompatible with solution or melt processing. The absence of solvent-based dewetting effects leads to pinhole-free, low roughness (<1 nm rms) ultrathin polymeric layers (<10 nm thin). Additionally, CVD can provide fully uniform conformal coverage, even over complex geometric features. Fabrication of devices on breathable textiles for wearable electronics and surface modification of porous membranes for improved separations processes are just two of many applications for conformal films.

As a result of dimensionless analysis and computation fluid dynamics, CVD polymer reactors have been scaled up to widths >1 m and from batch to semi-continuous roll-to-roll processing. These capabilities led to the founding of GVD Corporation, which currently manufactures robust CVD polymer coatings for an array of industrial applications. For example, GVD applies conformal low-surface energy CVD polymers to tire molds, allowing for the ready release of rubber from complex tread patterns during tire manufacture. More recently, DropWise Technologies Corporation was founded to commercialize robust ultrathin hydrophobic modification layers for improved condensation heat transfer.

567160 Enhancing Hydraulic Fracturing Productivity Via Model-Based Feedback Control

The term shale oil refers to natural oil trapped in rock formations of ultra-low permeability. Because of the low permeability of the shale formations, if a well were to be drilled into such a rock, it would take an extremely long time to reach an economically viable amount of production. What has made the recovery of shale oil economically viable, leading to the shale boom in the U.S. is the extensive use of two crucial technologies: hydraulic fracturing and directional drilling. Research on the relationship between proppant (i.e., propping agent) distribution and well performance indicates that uniformity of proppant bank height and suspended proppant concentration across the fracture at the end of pumping significantly affect the fluid conductivity, which is essentially the productivity, of fractured wells Specifically, in certain oil reservoirs, 10% difference in the propped fracture length from the desired value may cause 50% less oil production rate. However, no efforts have been made in improving the well performance via the regulation of the uniformity of proppant bank heights and suspended proppant concentration inside the fracture via real-time model-based feedback control of hydraulic fracturing, because of (a) large computational requirements due to the dynamic simulation of highly-coupled parabolic partial differential equations defined over time-dependent spatial domains, (b) spatial heterogeneity in rock mechanical properties such as the Young's modulus, and (c) limited availability of real-time measurements and presence of uncertainties in the measurement data. Motivated by these considerations, in this talk, I will introduce a systematic, computational modeling and control framework for real-time model-based feedback control of hydraulic fracturing.

559485 In Situ Grown Metal Nanoparticle Catalysts: Properties and Control

The ability to produce supported metal nanoparticle catalysts via exsolution from a perovskite type oxide support under reducing conditions has been known for some time and is used in some formulations of automotive emissions control catalysts. The ability to re-dissolve and exsolve the metal via redox cycling has led to these systems being referred to as smart or regenerable catalysts. While this phenomenon is well known, the mechanism by which the transition metal is exsolved from the oxide host is still poorly understood. The relationships between the exsolution process and the resulting structure of the metal nanoparticles are also not well understood. In this talk I will discuss our recent mechanistic studies of the exsolution process. Using well-defined model systems and detailed structural analysis by electron microscopy and atomic force microscopy, the nucleation and exsolution of Ni particles from Ni-doped strontium titanate was monitored. These studies show how exsolution produces unique surface structures consisting of metal particles partially submerged in pits on the oxide surface. I will show that this particle-in-a-pit morphology imparts unusually high thermal stability as well as significantly decreased carbon filament formation. Metal exsolution control will be discussed for host oxide lattices in thin film form.

564652 MFI-Type Zeolite Membrane Reactor for High Temperature Ethane Dehydrogenation

Ethene is the largest base chemical for the chemical industry and the increasing demand for ethene has stimulated substantial research into the development of new processes to reduce energy consumption. Catalytic ethane dehydrogenation using a packed bed membrane reactor (PBMR) is an attractive solution because the cracking equilibrium can be shifted in favor of ethylene by selective removal of hydrogen. In this study, we report the process intensification of ethane dehydrogenation reaction in PBMR operating with a Pt/Al_2O_3 catalyst. The effects of MFI zeolite membranes on ethane conversion and ethene selectivity are investigated. Increasing the reaction temperature and pressure helped in enhancing H_2 permeation across the membrane, which significantly increased the ethane conversion. It is demonstrated that medium-pore MFI membranes with moderate H_2/C_2H_6 selectivity can be effective in enhancing ethane conversion at high operation temperature by timely removal of H_2 through the membranes.

573875 Can Phosphorus-Doped Graphene be a Good Anode for Na-Ion Batteries?: A First-Principles Study

Secondary energy storage technologies based on sodium ions are an important emerging alternative to lithium ion batteries (LIBs). Sodium is less expensive than lithium, with its supplies being much wider and more democratically distributed throughout the globe. There are fundamental physical differences between Na and Li, and it is clear the knowledge available from extensive research on LIBs does not translate to sodium. However, the understanding of Na-based systems is at their infancy, and their complexities are just beginning to be understood. In this talk, we theoretically investigated the possibility of using phosphorus-doped (P-doped) graphene as an anode material in sodium ion batteries (NIBs). We reveal some fundamental physical properties of sodium adsorption on P-doped graphene by calculating formation and adsorption energies, voltage vs. capacity curves, migration for Na transport, and electron density of states and mobility. Our calculations suggest that Na adsorption on the same side of protrudent P is the preferred configuration for application in NIBs. This particular configuration possesses very large Na capacity (~582 mAh/g). More importantly, Na has to cross a diffusion barrier (~0.3 eV) and it diffuses in a zigzag path on the P-doped structures due to the presence of protrudent P with larger atomic size compared with the size of a carbon. We will also present electronic movements and accumulations during sodiation. Though carrier mobility may be not as fast as in metals or pristine graphene because P-doped graphene with Na adsorbed on it becomes a semiconductor, the calculated carrier mobility is still significantly large with the order of 10^3 to 10^5 $cm^2/V/s$. Lastly, we will discuss the guidelines for the efficient design of P-doped graphene as a promising anode for NIBs.

559480 Hydrogen Storage for PEM Fuel Cell Applications

Due to emerging economies and increasing population, demand for transportation fuels and energy is projected to increase rapidly. Currently, the majority of transportation fuels are derived from fossil fuels. Based on the current consumption rate, however, the supply of fossil fuels may last for <50 years. In addition, increasing CO_2 emissions due to burning fossil fuels put pressure on the ecological cycle and may account for global climate change. In this context, hydrogen is an attractive option to replace current hydrocarbon-based systems. A major obstacle for the development of hydrogen-powered fuel cell applications is the lack of safe, light weight and energy efficient means for on-board hydrogen storage. During the last twenty years, significant effort has been made to develop effective hydrogen storage methods, including hydrogen tank, sorbents, and metal/chemical hydrides. In this presentation, the current status of each onboard hydrogen storage method will be discussed, along with its advantages and disadvantages, and a perspective will be offered for future developments.

558372 Eradicating Antibiotic-Tolerant Bacteria and Biofilms By Antimicrobial Proteins Produced By Cell-Free Protein Synthesis

Chronic bacterial infections are difficult to treat through standard antibiotics because a small population of the bacteria in biofilms becomes tolerant to antibiotics by entering a dormant state. Such antibiotic tolerant bacteria are known as persisters and pose a significant problem in clinical settings. As antibiotic levels drop post-treatment, persister cells revert back into actively growing cells resulting in recalcitrant chronic infections. Currently, there are very limited approaches available for eradicating persister cells. Colicins, a type of antimicrobial bacteriocins, are considered as a viable alternative of conventional antibiotics due to their unique cell killing mechanisms that can damage cells by pore-forming on the cell membrane, nuclease activity, and cell wall synthesis inhibition. Additionally, colicins' cell killing activity is target-specific by recognizing receptors of the target cells without influencing other bacteria. Thus, colicins are excellent candidates for killing persisters. We utilized cell-free protein synthesis (CFPS) to produce colicins, as CFPS offers rapid protein production and characterization of such toxic proteins without cell-viability constraints and the need for protein purification. We demonstrated that colicins (*e.g.*, E1 and E2) produced by CFPS based on crude *Escherichia coli* lysates are active in eradicating persister cells induced by ciprofloxacin, by up to six orders of magnitude of cell viability reduction. Optimizing CFPS reactions such as including molecular chaperones and immunity proteins enabled significant improvement of solubility and activity of colicins. Colicin treatment was also effective in killing biofilm cells. Colicin production in the CFPS setting and cell-killing kinetics of cell-free synthetized colicins were very rapid, which provides a great potential for point-of-care drug producing systems by combining with lyophilized or paper-based CFPS. We further discuss the modular design of colicins that can redirect activity of colicins to target other bacteria. This study finds that colicins can eradicate antibiotic-tolerant bacteria and biofilm cells and that CFPS is a promising platform for rapid production, optimization, and characterization of toxic proteins.

560311 Hanwha Travel Award Flash Presentations

In this series of short "flash" presentations, the recipients of Hanwha Travel Award will provide snapshots of their ongoing research projects. It is expected that these presentations will introduce the promising young chemical engineers and their exciting research endeavors as well as enhance the ongoing interactions among industry, academia and chemical engineers with Korean heritage.

583199 Introduction

A concise overview of the 50th Anniversary of Computer Aides for Chemical Engineering (CACHE) Conference on "The Future of Cyber-Assisted Chemical Engineering Education," held in Breckenridge, Colorado, in July 2019 will be provided.

Dissemination plans for the conference outcomes include publication of several articles in the flagship magazine of AIChE, Chemical Engineering Progress, and a special issue of Chemical Engineering Education.

583200 CACHE History

The proposal founding the Computer Aids for Chemical Engineering Committee was submitted to the NSF on July 14, 1969 and first official meeting of the CACHE organization was held on November 18, 1969 in Washington DC. From its inception, CACHE has promoted the development and distribution of technology-based materials and software in chemical engineering education through projects applying computational chemical engineering, sponsoring conferences, recognizing outstanding contributions, and providing leadership in chemical engineering education. The purpose of this presentation is to review the accomplishments of CACHE over this 50-year period in supporting educational and research innovations.

583201 CACHE Initiatives: Systems Biology

In 2005, CACHE organized a new conference series entitled FOSBE (Foundations of Systems Biology in Engineering). The goal was to create a focused meeting on the topic of systems biology, as it relates to engineering methods and approaches, with a combined emphasis on: (i) applied problems from the pharmaceutical industry, (ii) research methodologies, and (iii) challenges in education and training.

The types of questions that were intended to be addressed by the conference included:
• How can one use systems biology and genomic data to analyze, interpret, and predict the relation between an organism's genotype and its phenotype?
• What are the best methods to combine data from hypothesis-driven research with data from high-throughput studies to create models of cells, communities of cells, and entire organisms? How can one drive iteration and innovation in this model building process?
• What is an appropriate computational infrastructure for maximizing the mining of bioinformatic data? What are the data format and databasing challenges?
• How do the complex network structures constrain intracellular signaling processing?
• What are the promising techniques for estimating model parameters from high throughput data records? How can identification methods be used to drive effective design of experiments?
• How can biological domain knowledge be combined with systems engineering methods to yield model reduction methodologies that capture essential features of biological regulation across multiple scales of time and space?
• How do we educate, train and develop the systems biologists of the future?

In this talk, I will report on the state of the FOSEB conference series, its unique position among other symposia and conferences, its collaborations with other professional societies, and the plans for the future.

583202 CACHE Initiatives: Molecular Modeling

The molecular modeling initiative in CACHE began in 1995 as its first foray into computational-based chemical engineering education beyond process design, operations and control. Molecular modeling has now been firmly established as one of the core competencies of CACHE, with around one quarter of the Board of Trustees representing this effort. In this talk, we will review the activities of CACHE in the field of molecular modeling, which includes educational initiatives and the establishment of the successful Foundations of Molecular Modeling and Simulation (FOMMS) conference series (fomms.org) that has been held every three years, beginning in 2000.

583203 CACHE Initiatives: Process Systems

This presentation describes CACHE initiatives in the area of process systems. CACHE supports the teaching of chemical engineering courses by managing webpages that link to resources that include syllabi, course schedules, computer-aided tools, interactive simulations, screencasts, concept questions, textbook information, useful links, and course notes. The resources relevant to the teaching of process systems include computer-aided educational tools developed by CACHE, such as sustainable manufacturing educational modules.

CACHE organizes conferences on progress and future directions in process systems, which include Chemical Process Control, Foundations of Computer Aided Process Operations, and Foundations of Computer-Aided Process Design. Activities and results are summarized for a newly founded conference on Foundations of Process Analytics and Machine Learning.

583204 Beyond A -> B: Computational Approaches for Education in Reaction Engineering and Kinetics of Complex Systems

Chemical reaction engineering (CRE) is a hallmark of the education of chemical engineers (ChE). It is often cited as a distinguishing feature a ChE compared to other engineers from other disciplines due to its clear marriage of chemistry and engineering. This presentation will focus on cyber-enabled tools that can be easily integrated into the undergraduate curriculum to introduce students to more complex reaction systems governing archetypal chemical processes, where an understanding of reactions well beyond A ® B is essential to guide process design and improvement of both novel and mature technologies. Automated tools for reaction network generation and approaches for automated model building will be discussed. Alternative solution methods for complex reaction systems based on stochastic simulations will be introduced, and software packages that are accessible for undergraduates to carry out kinetic Monte Carlo (kMC) simulations will be outlined. The inclusion of kMC as a core element of a CRE course will be motivated through examples of phenomena that are difficult to capture through analytical rate laws and how intuitive event-based simulations are to students.

583205 Graduate Research & Training in Molecular Data Science Through a University Wide Approach to Integrative Data Science Education

All areas of science and engineering have been the beneficiaries of a data-deluge. Chemical Engineering is at the forefront of this wave as a result of the commoditization of tools such as robotic instrumentation for molecular synthesis and characterization, ever increasing computational resources for simulation and design, high throughput experimental methods including gene sequencers, a proliferation of inexpensive and networked sensors and mixed modality data sources. To be competitive in the changing landscape of industry and academia, the next generation of chemical engineers will need to know how to effectively use data science tools such as statistical and machine learning, scalable data management and visualization techniques to communicate their findings effectively to stakeholders and peers.

The University of Washington (UW) has embraced data science in a campus wide initiative and the Chemical Engineering (ChemE) department at UW has led the effort to integrate highly contextualized molecular data science instruction at the graduate level, for both MS and PhD. ChemE @ UW offers two transcriptable options to graduate students, Advanced Data Science & Data Science. Students who participate in these options get a graduate degree in Chemical Engineering with a transcripted recognition of their participation in the program. The former is designed for chemical engineers who want to design and create new data science algorithms and tools, e.g. new machine learning methods for sparse molecular descriptor spaces, while the latter is intended for data science tool users who want to be confident in their selection of methods, e.g. choosing an effective neural network architecture for a predictive model of blood brain barrier permeability.

This presentation will review the structure of ChemE @ UW's graduate data science options from its inception as a result of two NSF training grants and in the context of the UW campus wide investment in data science education, research and training. It will highlight the NSF NRT Data Intensive Research Enabling Clean Technology (DIRECT) program which has been the primary vehicle for sustainable graduate molecular data science education in the department that includes 1) new graduate coursework – a sequence of flipped-classroom style courses that teach Data Science skills in a manner contextualized to molecules and 2) A capstone project that uses project based learning to practice and apply data science skills to challenging real world problems (in conjunction with external partners) in a team-based setting. It will close with outcomes from early cohorts that participated in the transcriptable options.

583206 Survey: Computing in Chemical Engineering

In 2003 the CACHE (Computer Aids for Chemical Engineering) Corporation carried out a survey of recent graduates in chemical engineering from four universities to determine how that group used (or did not use) computing in performing their job. Since that time, there have been considerable changes in the field of information technology, hence it is appropriate now to assess how the general conclusions reached fifteen years ago have changed. It is also valuable to have such information in order to consider changes in the curriculum, especially with respect to the use of and instruction about computing. Therefore we are conducting a similar survey to the one used in 1997 and 2003.

This survey is in two parts. One part is being emailed to chemical engineering faculty to assess what computing tools are currently being taught and used in the chemical engineering curriculum at universities. A second survey is being emailed to engineers in the chemical industry to assess computing in industry.

561599 Shape Control of Monodispersed Block Copolymer Particles

Shape and internal morphology of polymeric particles are of great importance for their unique physical properties. Self-assembly of block copolymers (BCPs) in evaporative emulsion provides a simple and effective route for the preparation of anisotropic particles with controlled shape and size. Understanding of thermodynamic phenomena associated with the bending/stretching of the BCP chains confined within the particles is necessary to enable precise control of the shape and microstructure of the particles. Herein, we report the systematic design of shape-anisotropic diblock copolymer (dBCP) particles based on a new theoretical model that includes entropic penalty associated with bending of dBCP chains upon deformation of the particles. First, we produced convex-lens shaped (oblate) and football-shaped (prolate) PS-*b*-PDMS particles, where the aspect ratios (AR, defined as the major axis divided by minor axis) were varied. Of note, the AR of the oblate particles increased almost linearly up to 10 as the particle size increased, whereas the increase of AR for the prolate particles was limited to 2.0. For oblate particles, the high bending energy of the curved cylinders at the periphery of a particle can be released by increasing the AR of particle. However, the relatively low bending energy of curved lamellae of prolate particles prevents the particles from having a high AR. Furthermore, our theoretical model that considers these bending energies successfully explains the experimental observations on the variation of particle shape depending on particle size and the dBCP molecular weight.

In second part of my talk, I will talk about the use of these shape-controlled particles for the development of the colorimetric sensing system. As an example, I will present the development of dual-responsive shape-switchable BCP particles, which respond to subtle temperature and pH changes near physiological conditions (*i.e.* human body temperature and neutral pH). The shape transition of PS-*b*-P4VP BCP particles between lens and football shape occurs in very narrow temperature and pH ranges: no temperature-based transition for pH 6.0, 40-50 °C for pH 6.5, and 25-35 °C for pH 7.0. To achieve these shape transitions, temperature/pH-responsive polymer surfactants of poly(DEAEAM-*r*-NIPAM) are designed to induce dramatic changes in relative solubility and their location in response to temperature and pH changes near physiological conditions. In addition, the BCP particles exhibit reversible shape-transforming behavior according to orthogonal temperature and pH changes. Colorimetric measurements of temperature and pH changes are enabled by shape-transforming properties combined with selective positioning of dyes, suggesting promising potential for these particles in clinical and biomedical applications.

References

[1] B. J. Kim* et al. ***Macromolecules*** 2019, 52, 1150

[2] B. J. Kim* et al. ***Adv. Funct. Mater.*** 2018, 28, 1802961 (Invited review)

[3] B. J. Kim* et al. ***Advanced Materials*** 2017, 29, 1700608

[4] B. J. Kim* et al. ***ACS Nano*** 2019, **DOI:** 10.1021/acsnano.8b09276

[5] B. J. Kim* et al. ***ACS Nano*** 2017, 11, 2133

564221 When Misbehaving Proteins Met Well-Behaved Proteins…

Self-assembly of intrinsically disordered polypeptides (IDP) imparts toxic biological effects depending on the size in over 20 neurodegenerative diseases. Unfortunately, our understanding of amyloid polypeptides, as related to biomedical implications, is limited by their self-assembling nature. In this talk, I will first present our study on the creation of a dual peptide system, where a pair of β-amyloid (Aβ) variants, KLVFWAK and ELVFWAE, are not self-assembled but hetero-assembled in the presence of their assembly partners. We provide evidence that the resulting hetero-assemblies share molecular, structural and morphological similarities with typical amyloid self-assemblies formed by a single polypeptide (e.g., Aβ). We anticipate that our dual peptide system may readily be adapted for precise control of amyloid assembly, for the study of size-dependent neurotoxicity.

Unlike an IDP, a globular protein folds into a compact structure, responsible for various biological functions. Interestingly, globular proteins are also known to form amyloid fibrils under partially denaturing conditions, demonstrating that amyloid aggregation is a generic property of proteins. In the second half of this talk, I will present our study where a globular protein, *Bacillus circulans* xylanase (BCX), can aggregate to form amyloid fibrils under native conditions. Interestingly, addition of KLVFWAK or ELVFWAE modulated the BCX amyloid aggregation. This study also provides insight into a correlation between the kinetic stability and amyloid aggregation of BCX, and supports a view that Aβ-derived fragments can be useful for the modulating amyloid aggregation of some, though not all, globular proteins.

564247 Polymer Lung Surfactant Therapy for Respiratory Distress Syndrome

Animal-derived lung surfactants annually save 40,000 infants with neonatal respiratory distress syndrome (NRDS) in the US. Lung surfactants have further potential in treating nearly 200,000 adult patients with acute respiratory distress syndrome (ARDS) each year in the US. ARDS is a complication associated with increased alveolar membrane permeability followed by subsequent lung surfactant deactivation and inflammation in the lungs. Often described by patients as "like drowning from within", lack of functioning surfactant causes alveolar collapse, thus resulting in impaired oxygenation of the blood. This problem aggressively propagates into multiple organ failure and eventually death for about 40% of patients. All prior clinical trials testing NRDS therapeutics (animal-derived lung surfactants) in ARDS patients were unsuccessful. For use in treatment of ARDS, the properties of current NRDS therapeutics need to be modified. Although the limitations of current therapeutics have been known since 1990s, there has not been much improvement. As a result, there are currently no existing therapeutics for ARDS, and treatment remains limited to invasive options such as mechanical ventilation and ECMO. To address this need, our laboratory has been exploring a radically different approach, in which, instead of conventional lipid and protein-based formulations, synthetic biocompatible polymers are used as the active therapeutic ingredient. This talk will discuss our endeavor, driven by engineering curiosity and a humanitarian motivation, to develop a first-in-class polymer lung surfactant therapy for ARDS treatment.

567731 Additives Applied to Cu Electrodeposition

Copper (Cu) is widely used as a metallization material in semiconductor fabrication because of its properties like a low electrical resistivity. As the one of the methods for Cu deposition, electroplating has been extensively used for the high quality copper on a substrate. The mechanical properties of the electrodeposited Cu can be controlled with the modification of current/potential profiles such as pulse and pulse-reverse deposition, or the use of additives. Especially, additives can control the properties maintaining electroplating process, and also enable to modify the deposition rate topologically. Various combinations of additives are used to form the deposit with suitable properties and induce bottom-up filling in various scales of vias and interconnects. Conventionally the additives for Cu electrodeposition are classified into three categories based on the functionality. Accelerators increase the deposition rate of Cu where they adsorb. Instead, suppressors retard the Cu electrodeposition rate, physically blocking the access of cupric ions to the surface of the electrode. Levelers, generally slowing the deposition rate, similar to suppressors, have specific behavior dependent on the convective condition, and are introduced to fill micro-scaled feature and through-holes.

Through silicon via (TSV), the focused technologies for 3-dimensional (3D) packaging to solve the scaling limitation, is one of the fields using Cu electrodeposition. In the Cu deposition for fabrication of TSV interconnect, leveler acts as key materials, which is essential to maintain the inhibition at the top of TSV due to its large scaled structure and long filling time compared to damascene features. Therefore, a number of studies related to the levelers have been introduced to obtain bottom-up filling of micro-scaled features and reduce process time.

Our group also has been dedicated to study about additives applied to Cu electrodeposition, especially levelers.

New organic compounds have been developed as levelers for feature filling. This compound containing ammonium cation and halide ions strongly suppressed electrodeposition on the side and top of TSVs. These choline-based levelers induced selective deposition at the bottom of TSVs and established a flat growing surface, which can facilitate bottom-up filling of TSVs without any defects. The mechanism of gap filling is demonstrated based on the electrochemical analyses and filling profiles.

Furthermore, iodide ion which is the inorganic material has been introduced, which can function like organic leveler. It inhibited Cu deposition convection-dependently. The inhibition mechanism of iodide ion was identified based on chemical and electrochemical reaction in solution chemistry containing Cu ions, iodide ions, and other additives. This inorganic material formed non-conducting material, CuI, as inhibition layer on the Cu surface. Its convection-dependent behavior was suitable for bottom-up filling of TSVs.

CuI formed during electrodeposition with iodide ions could affect reliability of filling solution as byproduct. To solve this issue, another inorganic leveler was studied, bromide ion. The bromide ion cannot inhibit Cu electrodeposition as CuBr form due to its lower reduction potential. Although it showed the weaker inhibition strength than iodide ions, the synergistic effect with PEG-PPG enables to fill the TSV without any defects. Moreover, the filling rate doubled with this ion because of higher deposition efficiency than choline-based organic leveler or iodide ion. The mechanism of bromide ions was investigated by comparing to that of iodide ions.

In addition to effect of organic/inorganic additives on Cu electrodeposition, we have investigated the degradation mechanism of additives. As the Cu electrodeposition proceeds, the process accompanies the degradation of the plating solution, as like other chemical processes. Plating solution contains many kinds of components such as Cu metal salts and additives, and bath conditions can be changed with degradation of these materials. Consumption of the metal source is easily complemented by supplementing the additional metal salt. However, unlike Cu metal ions, additives are degraded with forming byproducts during electrodeposition. The formation of byproduct could adversely affect to reliability of the Cu electroplating process, causing defects such as voids and seam in interconnect, and unexpected change of Cu properties. Therefore, the degradation mechanism of each additive should be identified to understand

and control the solution chemistry. The mechanism of bis(3-sulfopropyl) disulfide (SPS) and polyethylene glycol-polypropylene glycol copolymer (PEG-PPG) were investigated by electrochemical analysis and spectroscopic method.

SPS, widely used as an accelerator, is degraded easily in the electroplating process. SPS is decomposed into 3-mercapto-1-propanesulfonic acid (MPS) or intermediate oxide forms during the electrodeposition, and lastly converted into 1, 3-propane disulfonic acid (PDS). The formed MPS is reduced to SPS through interconversion reaction, and, on the other hand, PDS is irreversibly accumulated in the electrolyte.

The degradation reactions, which occur at both Cu cathode and insoluble anode, were confirmed electrochemically and chemically, using a membrane to separate each electrode. At the insoluble anode, SPS is decomposed mostly into PDS through irreversible electrochemical reaction. However, SPS is also decomposed by a chemical reaction as well as the electrochemical reaction at the Cu cathode. Cupric ions convert into cuprous ions by disproportionation reaction with Cu electrode, and these cuprous ions forms free radicals with dissolved O_2 in the solution. SPS and Cu(I)-MPS react with the free radicals, and finally converted into PDS.

PEG and PEG-PPG copolymer, which are typically used suppressor, are slowly decomposed compared to the accelerator. In the case of this suppressor, it is divided into the small molecular weight of PEG. As the electrodeposition proceeds, the molecular weight of the PEG decreases to less than a certain value, not showing suppression effect. Moreover, by degradation of PEG, byproducts are produced, having a functional group such as hydroxyl (-OH), aldehyde (-COOH), formic ester (-COO), ketone (C=O), and propyl ($-C_3H_6-$).

Because of these degradation reactions, monitoring the condition of plating bath is necessary for reliability of processes, which has been optimized by our group.

It is difficult to monitor the change of additives' chemical properties. Therefore, monitoring the additives concentration can be alternative to monitor the condition of the solution. Additives are decomposed to byproduct and incorporated into the deposit due to their chemical/electrochemical effects. Therefore, the degradation of additives decreases the additives' concentration. Because the electrochemical behavior of the additives is critically affected by the concentration, it can be measured using cyclic voltammetry stripping (CVS).

CVS method uses the voltammetric approach in determining the concentrations, and it focuses on the stripping charge, which represents the electroplating rate influenced by the type and concentration of additives. Calibration curve of a target additive is obtained by CVS using standard solution whose concentration of the target additive is known. Using this curve, the concentration of the additive by interpolating the stripping charge of a sample solution can be determined.

For monitoring additives used in Cu electrodeposition solution, CVS methods have been optimized for each type of additives. Modified linear approximation technique-CVS (MLAT-CVS) is used for determining concentration of accelerator. However, MPS, the degradation byproduct, acts as an interference due to its different accelerating effects on Cu electroplating compared with SPS. Therefore, the function has been devised related to the conversion ratio from SPS to MPS, which enabled the determination of individual SPS/MPS concentration. Moreover, by controlling the pH of the base and target solutions, MPS was completely converted to SPS, and the total amount of accelerating agent could be determined when using MLAT-CVS method.

Dilution/titration-CVS (DT-CVS) method is generally used for determining concentration of suppressor. When the conventional DT-CVS analysis was applied to two-additive target solutions containing both accelerator and suppressor, the accelerator could affect stripping charge as an interference. Even small concentrations of accelerator in CVS bath can increase error in determined suppressor concentration. We suggested new method of determining the suppressor concentration in two-additive solution by introducing iodide ion which completely suppressed effect of the accelerator.

As mentioned above, iodide ion has been recently developed for defect-free TSV filling due to its excellent inhibition effect by deactivating SPS. Response curve-CVS (RC-CVS) method could be applied to monitor concentration of iodide ion due to its electrochemical behavior. Optimizing the determination process based on the mechanism of I^- affecting Cu electrodeposition, the successful monitoring of I^- concentration was performed in wide range with high resolution. This method was used to recover the performance of the electroplating solution containing three additives for TSV filling.

577371 Multiscale Modeling of Pulp and Paper Manufacturing Process

In chemical pulping process, the cell wall thickness (CWT) of wood chips significantly affects paper properties such as bending stiffness, tensile strength, tear strength, ink holdout, flexibility, surface smoothness, etc; consequently, it will decide whether paper is suitable for printing or packaging. In this talk, I will present a novel multiscale model by combining the mass continuity and thermal energy balance equations adopted from a modified "extended Purdue model" with a kinetic Monte Carlo algorithm to describe the microscopic events such as the evolution of CWT and Kappa number; otherwise, these microscopic properties of wood chips are not accessible by existing measurement techniques. To reduce the model complexity, a data-driven reduced-order model is developed, which is then used for developing a soft sensing system (i.e., Kalman filter) to estimate the primary measurements (e.g., CWT, Kappa number) by utilizing the secondary measurements (e.g., active effective alkali and dissolved lignin concentrations in the free-liquor phase). Based on this soft sensing system, a model-based feedback control framework is developed to regulate both the CWT and Kappa number of wood chips.

572201 Tissue-Engineered Trabecular Bone Model Using Demineralized Bone Slices

The trabecular bone is a multi-functional dynamic tissue that maintains mechanical, mineral, and blood homeostasis in response to continually changing external physical stress and internal physiological need, respectively. The complexity of trabecular bone tissue has been increasingly classified mainly using mouse models, but in vitro reconstitution of trabecular bone tissue complexity and metabolism has remained challenging due to the lack of relevant experimental models. Remodeling sites collectively cover about 5% of the trabecular bone surface at any time, and the rest surface remains quiescence. Bone-forming osteoblasts and bone-resorbing osteoclasts drive the bone remodeling process. Accumulating data indicate that osteocytes, the primary mechanosensory cells residing beneath of the lamella bone play a central role in regulating the bone surface metabolic activity via secreting a set of soluble factors. In this talk, I will describe our on-going effort to create in vitro functional trabecular bone tissue models that represent both cellular and extracellular complexity in a controlled and analytical manner.

Many investigators believe that bone is probably the best biomaterial for creating bone tissue. We developed a method to process a compact bovine bone into demineralized thin slices (20μm). We hypothesize that multiple layering osteoblast pre-seeded demineralized bone slices will reproduce the lamellar structure of bone tissue complexity. Demineralized bovine bone slices support osteoblast differentiation, further maturation into osteocytes, and osteoclasts. When osteoblasts pre-seeded demineralized bone slices are stacked and exposed to cyclic compression under hypoxic milieu, they differentiate into osteocytes and reproduce both surface and subsurface trabecular bone tissue complexity. The prepared resting state of trabecular bone model is chemically stimulated when exposed to vitamin D3 and prostaglandin E2 with the substantial shift of bone remodeling cytokine profiles such as OPG and RANKL. We envision that our tissue-engineered trabecular bone models with demineralized bone paper will allow understanding various aspects of trabecular bone biology and harnessing their intrinsic functions for clinical applications.

559572 Infiltration of Polymers into Disordered Nanoparticle Packings: Polymers Under Extreme Nanoconfinement

Capillary rise infiltration (CaRI) is a new method of fabricating polymer nanocomposites with extremely high fractions of nanoparticles (> 50 vol%). In CaRI, highly filled composites are formed by thermally annealing a bilayer of polymer and nanoparticle, which induces imbibition of polymer into the interstices of the nanoparticle packing. The chain dimension of the polymer, which depends on its molecular weight, can be comparable to or greater than the average pore size of the nanoparticle packing; thus this system provides a powerful platform to study the effects of extreme nanoconfinement on the transport and thermomechanical properties of polymers. I will share our current understanding of the transport phenomena and glass transition of polymers under such nanoconfinement using a combination of experimental and computational approaches.

579938 Flowing Complex Fluids, from Blood to the Buffer Layer

Over the course of his distinguished career, Bill Schowalter made important contributions to understanding the dynamics of multiphase and non-Newtonian flows. In his honor, these will be the topics of this inaugural William R. Schowalter Lecture. The first part of this talk will focus on a multiphase flow that is important to all of us, the flow of blood in the microcirculation. Red blood cells migrate toward the center of a blood vessel, leaving a cell-free layer at the vessel wall, while white blood cells and platelets are preferentially found near the walls, a phenomenon called margination. A mechanistic theory has been developed to describe segregation in blood and other confined multicomponent suspensions. The theory predicts, in good agreement with direct simulations and experiments, that the cell-free layer thickness follows a master curve with confinement ratio and volume fraction. It also predicts several regimes of segregation, depending on the value of a new quantity called the ``margination parameter". Experiments performed in the laboratory of Wilbur Lam indicate the physiological and clinical importance of these observations.

The second part of this talk addresses one of the most puzzling and dramatic phenomena in non-Newtonian flow, the substantial reduction of turbulent energy dissipation (drag) that occurs when a small amount of a long-chain polymer is added to a liquid. If the Reynolds number is sufficiently low, a turbulent channel flow will laminarize as polymer concentration increases, but then become turbulent again at higher concentration. Direct simulations of flow in the latter regime, along with linear analyses performed by Beverley McKeon's group, yield a surprising result: the turbulent fluctuations strongly resemble so-called Tollmien-Schlichting (TS) waves. Although these are found in Newtonian flows, they do not play a role in fully-developed Newtonian turbulence. In the polymeric case, however, polymer stresses suppress the normal turbulent structures while amplifying the TS modes. The emergence in viscoelastic flow of flow structures with Newtonian origin may shed light on the observed near-universality of the polymer drag reduction phenomenon with regard to polymer properties.

579709 Opening Comments for Session on Thermodynamics Needs of the Chemical Industry

A brief review of the history and objectives of AIChE's session on Thermodynamics Needs of the Chemical Industry.

554065 Current State of Thermophysical Property Measurements and Reporting in Scientific Publications

Experimental thermophysical and thermochemical property data for chemical substances, mixtures, and materials are needed in various fields, such as chemical engineering, design of new processes and products, development of predictions/correlations, and education. Two factors are critical for addressing the above needs: availability and quality. Depending on their quality, the data can be useful or misleading.

While the amount of published thermophysical data constantly increases and their searchability improves, the reliability of the data in each case requires exploration. A set of methods and tools for data quality analysis is used at the NIST Thermodynamics Research Center (TRC) [1]. Application of those methods reveals the extent of inconsistencies and provides statistical information. For the journals cooperating with NIST/TRC for data validation, we apply our checks to accepted manuscripts and can compare their state before and after interaction with TRC. The aspects we consider are completeness of data reporting, physical/thermodynamic validity of data, and numerical representation of data. A significant part (up to 50%) of manuscripts has fixable problems such as incomplete reporting, ambiguities in identification of substances and properties, and typographical errors. Non-fixable problems such as flaws in technique, non-physical or inconsistent values are also frequently observed. The statistics of both will be provided in the presentation. While problems detected during the journal cooperation process are typically fixed after TRC provides a data report to the journals prior to publication, the rate of data problems in other journals is generally higher. Hence, no published property value can be blindly used without expert evaluation.

Part of TRC's activities are dedicated to improvement of the interaction with the cooperating journals. The experience gained in that process is used more broadly with the intent to increase the efficiency of thermophysical research and better serve the needs of data users. An element of those activities is the Good Reporting Practice initiative. Other planned actions will be dedicated to assisting reviewers and experimentalists with a goal to obtain more reliable results from different kinds of measurements.

This presentation will provide the statistics in more details, including an analysis of possible causes of the observed measurement and publication mistakes. It will also mention a potential effect of data errors on process simulation.

[1] Vladimir Diky, Ala Bazyleva, Eugene Paulechka, Joseph W. Magee, Vikina Martinez, Demian Riccardi, Kenneth Kroenlein. Validation of thermophysical data for scientific and engineering applications. J. Chem. Thermodyn, 2019, 133, 208-222.

*Corresponding author: *Email: ala.bazyleva@nist.gov;* Phone: +1-303-497-5981

569240 An Improved Spectroscopic Method for Determination of Association/Solvation Parameters Used in Process Models

Process modeling of systems continues to be challenging when association and solvation are present. Accurate modeling of mixtures with such species will become increasingly important because renewable chemicals, fuel intermediates and biopolymers typically associate via hydrogen bonding. Equations of state such as SAFT, CPA and ESD have gained popularity by explicitly representing hydrogen bonding. In a previous study, we have developed an activity-coefficient thermodynamic model which incorporates Wertheim's perturbation theory (TPT-1). However, association parameters used in literature differ significantly depending on the type of data used to derive them. For example, the parameter for the self-association of methanol can vary by up to a factor of four.

This work focuses on the determination of the Wertheim association parameters using spectroscopic measurements such as FTIR. While this approach is not new, it is complicated by the challenges involved in quantifying the associating clusters such as alcohol monomers and dimers. Issues persist in curve fitting for evaluating the spectra where absorbances are highly concentration and temperature dependent. More importantly, the significant variation in the IR molar extinction coefficient across the broad hydroxyl band has been ignored in most studies and thus hindered accurate quantification of association. To address these challenges, we have developed a method based on quantum calculations for scaling experimental FTIR spectra such that more meaningful extraction of model-relevant thermodynamic constants is possible. The parameters acquired from our methods are compared to those obtained from other studies and used to model the phase equilibria of binary alcohol-alkane mixtures. The results demonstrate the capabilities of the activity coefficient model and further prove the importance of accurately capturing the hydrogen bonding phenomena in thermodynamic modeling.

570182 A Molecular-Based Screening Tool for Water-Free Amine Solvents for CO_2 Capture

Currently, the chemical absorption of CO_2 using aqueous solutions of amines mainly, is the most mature technology for CO_2 capture and separation, that has been applied in many industrial processes. However, the high regeneration energy requirement is the major hurdle to the large-scale deployment of this technology. Among the various strategies present as a remedy for this issue, the utilization of water-lean or water-free solvents provide an attractive alternative. This is done through either partial (water-lean) or full (water-free) replacement of water with alternative physical solvents with lower heat capacity and evaporation enthalpy that, when combined with traditional alkanolamines, form physical-chemical absorbents. The accurate determination of CO_2 solubility in the utilized solvents is an integral factor in the design and simulation of carbon capture or acid gas treatment processes along with their techno-economic evaluation. However, the lack of experimental data on the thermophysical properties of new amine-based solvents or blends is a major obstacle for reliable process design and simulation. Therefore, devising a consistent and reliable thermodynamic model capable of accurately describing the phase equilibria, thermophysical and transport properties of CO_2 and amine-based systems is highly needed.

In this contribution, we demonstrate the capabilities of a robust molecular-based framework for the description of key thermophysical properties of water-free and water-lean amine-based solvents for the efficient removal of CO_2 and other acid gases from industrial gas streams at relevant gas separation process conditions. The screening tool is built on the molecular-based equation of state (EoS) soft-SAFT [1] combined with the Free-Volume Theory for the integrated modeling of phase behavior, enthalpies, densities and viscosities. Solvents investigated in this work include mixtures of various alkanolamines with different physical solvents such as alcohols, glymes, glycols, polar aprotic solvents and ionic liquids (ILs). Molecular parameters of all the investigated substances are either taken from previous contributions or developed in this work and used here in a transferable manner for describing the solubility of CO_2 in the selected solvents at conditions of relevance for industrial applications.

The description of the chemisorption of CO_2 in the studied amine + physical solvent blends is done in a consistent manner with that for the chemisorption of CO_2 in aqueous amine reported in our previous contributions [2,3]. A scheme of implicit reactions is used to describe the formation of carbamate and/or bicarbonate products resulting from the chemical reactions between CO_2 and the examined amines. The influence of the organic solvents on the chemical reactivity of amines with CO_2 in water-free solvents is investigated in detail. This procedure eliminates the need to specify the detailed equilibrium reactions and significantly reduces the number of parameters required to represent the absorption process. A maximum of two adjustable model parameters, optimized for a fixed amine concentration, suffice to represent the absorption of CO_2 in the studied solvents over a broad range of conditions, with high predictive and extrapolative capabilities. The acquired results from our the developed thermodynamic framework are compared to experimental data gathered from literature and predictions from the recommended thermodynamic models implemented in the Aspen Plus® (V.10) process simulator and the model found to be capable of providing reliable solubility estimations over a broad range of pressure, temperature and compositions.

The collection of these results demonstrates that once the robust molecular model is developed and validated, it can be used as a predictive tool, combined with macroscopic thermodynamics, as a reliable platform for the screening of chemical solvents as function of key process parameters, and as a valuable tool for process modelling and optimization.

This work is partially funded by the ADNOC Gas Processing and its stakeholders through the Gas Research Center (project GRC2018-003) and Khalifa University of Science and Technology (RC2-2019-007).

References

[1] F.J. Blas, L.F. Vega, *Mol. Phys*. 92, 135–150 (**1997**). F.J. Blas, L.F. Vega, *Ind. Eng. Chem. Res*. 37, 660–674 (**1998**).

[2] J.O. Lloret, L.F. Vega, F. Llovell, *J. CO_2 Util*. 21, 521–33 (**2017**).

[3] L.M.C. Pereira, F. Llovell, L.F. Vega, *Applied Energy*. 222, 687-703 (**2018**).

571049 Relationship of Solubility and Thermodynamic Parameters for Amine-Containing Non-Regenerative H_2S Scavengers

Dan F. Smith Department of Chemical Engineering, Lamar University, P.O. Box 10053, Beaumont, Texas 77710, United States

Removal of H_2S from oil and gas streams has been a priority, oftentimes a necessity, for upstream crude oil/gas facilities for decades. However, the advances in production, namely horizontal drilling and hydraulic fracking, have produced oils and natural gases that contain higher concentrations of H_2S than previously seen. Crudes containing > 0.5 % sulfur are considered sour and must undergo separation and/or conversion processes to lower the H_2S content prior to entering the midstream pipelines and railcars. H_2S scavengers can be categorized into two types: regenerative and non-regenerative. Regenerative scavengers, such as alkanolamines, absorb H_2S using reversible absorption with stripping recovery processes. With non-regenerative scavengers, an irreversible reaction converts the H_2S to form sulfur-containing hydrocarbons. Non-regenerative scavengers are usually aldehydes, ethers, and amines (i.e. acrolein, glyoxal, formaldehyde, alpha-aminoethers, or triazines). Triazine-based scavengers have become the most widely used of the reactive-type scavengers. 1,3,5-hexahydrotriazine is the most frequently used; however, its byproducts equipment fouling, particularly overhead condensers within refineries.

These harmful effects have led to molecular adaptations to the basic triazine molecule whose byproducts would be benign to equipment fouling and corrosion. Varying side chains have the impact of making the triazine scavenger more (or less) soluble in aqueous (or non-polar) crudes. In addition, these side chains could be tailored to be more active for sour gas, rather than sour liquid crude streams. Side chain modifications lead to uncertainty of solubility, liquid/gas interactions, and possible byproduct complications. In this work, solubility and scavenger/ion interactions have been determined for triazines containing ethanolic, amino, bromine, and acrolein substituents, namely 1,3,5-Triacryloylhexahydro-1,3,5-triazine, cyanuric acid, and 1,3,5-Tris(2-hydroxyethyl)cyanuric acid. By adding salts to the mixture, the solubility data yields insightful information towards the ionic mediums in which these materials could be used. Moreover, Setchenow's coefficient, the octanol/water partition coefficients, and the activity coefficients have been determined to examine the potential fate these scavengers when considering environmental factors. The experimental results and model development will be presented.

558709 Minimizing Process Development Costs in Industry By Computational Molecular Modeling Screening: Separation of Enantiomers for Pharmaceutical Applications

The overall goal of this project is to develop an efficient (time and cost) and reliable computational molecular modeling for screening conditions for effective separation of enantiomers using chiral stationery phases. More than half of all pharmaceuticals are chiral compounds. The potential market for enantiomer separation is $1.2 billion. Although enantiomers of chiral compounds have the same chemical structure, they can exhibit marked differences in physiological activity; therefore, it is important to remove the undesirable enantiomer. Chromatographic separation of chiral enantiomers is one of the best available methods to obtain enantio-pure substances, but the optimization of the experimental conditions can be very time-consuming. Generally, there are several chiral stationary phases to choose from; the most commonly used ones are polymers that have an amylose or cellulose backbone, with various sidechains, coated or covalently bonded to a solid support. Then there is the choice of the solvent system that constitutes the mobile phase. Furthermore, modifiers (acids or bases or buffers) may be added to the solvent in order to change or protect some functional groups on the drug to effect a better separation between the enantiomers. One could try different CSPs and mobile phases using educated guesses, but the number of possible combinations is huge, and could easily run into hundreds of combinations. This has stymied progress for developing new drugs especially orphan drugs. To overcome these problems, we have used molecular dynamics simulations to screen for these choices. Chromatographic enantiomer separation is a dynamic process, with the interactions between the drug and the chiral stationary phase mediated by the solvent, thus, no single interacting structure, such as could be found by finding low-energy structures via MD or quantum calculations, could possibly describe and account for the ratio of residence times in the chromatographic column for the enantiomeric pair. Atomistic molecular dynamics simulations permit us to develop an approach to predict which enantiomer elutes first and also provide an estimate of the separation factor.

AUTHOR INDEX

AUTHOR INDEX